I0055233

# Psychoactive Plants

## *Ethical Issues and Basic Evaluations*

**James O. Fajemiroye, Ph.D.,** was born and brought up in a community where medicinal plants usage was a daily experience. His formal education was at the University of Ilorin, Nigeria where he earned a B.Sc. in Plant Biology (formally referred to as Botany). He conducted ethnobotanical survey of Ilorin Emirate while working as industrial trainee under a renowned horticulturist, Mr Anthony Okulaja. He furthered his education by taking an M.Sc. (Area of specialization - Cellular and Molecular Biology) and Ph.D. in Biological Sciences (Area of specialization Pharmacology and Physiology) under the supervision of Dr. Elson Alves Costa at the Federal University of Goias, Institute of Biological Sciences in 2012 and 2015, respectively. His researches focused on preclinical evaluation of crude extracts and/or phytoconstituents with anxiolytic and/or antidepressant properties. He also worked under Dr. Jordan K. Zjawiony, who was the co-advisor of his Ph.D. thesis, at the University of Mississippi, School of Pharmacy. His area of research included the pharmacological evaluation of psychoactive plants, compounds (natural or synthetic) and chemical modification of biologically active natural products. He was awarded a prestigious 2-year postdoctoral fellow (*Programa Nacional de Pós Doutorado*) to develop analytical method for pharmacokinetic profiling of behavioural alterations in experimental animals. The latter brought his pharmacological research experience full circle. He was able to complement pharmacodynamics aspect of natural product with the pharmacokinetic and toxicological studies. He acted as a student representative on the Board of International Society for Ethnopharmacology (ISE) and just recently, he was elected as a full Board member of ISE. He is a teacher, editorial board member, reviewer, examiner, keynote speaker, co-author of book and articles.

# Psychoactive Plants
## Ethical Issues and Basic Evaluations

– Editor –

**James O. Fajemiroye (PhD)**

– Contributors –

**Roberto Saavedra (MSc)**

**Dr. Gustavo Rodrigues Pedrino**

**Lara Marques Naves (MSc)**

**Dr. Luiz Carlos da Cunha**

**Aderoju Adeleke Adesiyan (MSc)**

**Dr. Heleno Dias Ferreira**

**Dr. Christianah Abimbola Elusiyan**

**Dr. Elson Alves Costa**

# Kruger Brentt
## P u b l i s h e r s
### 2 0 1 7

Kruger Brentt Publishers UK. LTD.
Company Number 9728962

Regd. Office: 68 St Margarets Road, Edgware, Middlesex HA8 9UU

© 2017 Editor

*Disclaimer:*

*Every possible effort has been made to ensure that the information contained in this book is accurate at the time of going to press, and the publisher and author cannot accept responsibility for any errors or omissions, however caused. No responsibility for loss or damage occasioned to any person acting, or refraining from action, as a result of the material in this publication can be accepted by the editor, the publisher or the author. The Publisher is not associated with any product or vendor mentioned in the book. The contents of this work are intended to further general scientific research, understanding and discussion only. Readers should consult with a specialist where appropriate.*

*Every effort has been made to trace the owners of copyright material used in this book, if any. The author and the publisher will be grateful for any omission brought to their notice for acknowledgement in the future editions of the book.*

*All Rights reserved under International Copyright Conventions. No part of this publication may be reproduced, stored in a retrieval system, or transmitted in any form or by any means, electronic, mechanical, photocopying, recording or otherwise without the prior written consent of the publisher and the copyright owner.*

Library of Congress Cataloging-in-Publication Data

Psychoactive plants : ethical issues and basic evaluations / editor, James O. Fajemiroye (PhD) ; contributors, Roberto Saavedra (MSc.) [and seven others].
    pages cm
    Contributed articles.
    Includes bibliographical references and index.
    ISBN 978-1-78715-000-3 (hardbound)

    1. Psychotropic plants. 2. Medicinal plants. 3. Psychotropic plants--Moral and ethical aspects. 4. Medicinal plants--Moral and ethical aspects. I. Fajemiroye, James O., editor.
    QK99.A1P79 2017      DDC 581.634    23

For information on all our publications visit our website at http://krugerbrentt.com/

# Acknowledgement

Special appreciation to the *Universidade Federal de Goiás, Programa de Pós-Graduação em Ciências Biológicas, Programa de Pós-Graduação em Ciências Farmacêuticas,* National Centre for Natural Products Research, *Coordenação de Aperfeiçoamento de Pessoal de Nível Superior, Conselho Nacional de Desenvolvimento Científico e Tecnológico, Fundação de Amparo à Pesquisa do Estado de Goiás,* family and friends.

*James O. Fajemiroye*

# Foreword

The interaction between human and plant has been a rich source of information in medicinal plants research for years. Medicinal plants remain a leading source of compounds with therapeutic value. The researches involving these plants have changed the world of medicine. The chapters in this book are well organized in such a way that reflect the enthusiasm of the authors in convey the details of basic research as it relates to medicinal plants. The book ignites curiosity, excitement and motivates high school students, undergraduate students and young scientists into ever expanding multidiscipline field of basic science.

This book collates scientific names of medicinal plants, active principles of plant origin with intricate structure, popular indications of different species, and preclinical evaluations with their respective mechanism of actions. The authors have included hundreds of medicinal plant species with several phytoconstituents in the book to showcase the importance of this field of research to new drug discovery. An abundance of colorful pictures and illustrations make the book an easy read. The interesting histories in the book connect procedural steps, chemistry and biology in a very appealing way. The book offers opportunity for readers that are seeking to get the "know-how" of bench works and broader perspective of basic research. Although the emphasis of this work is on medicinal plants, it contains much that will be of interest to non-scientists, researchers from different field that are interested in scientific procedures, ethics and safety measures in basic science.

It is also a unique opportunity for the reader to take advantage of relevant references in the book to revisit interesting topics. As this captivating book takes readers through history, ethics, field and bench screening approaches, it is important to state that the lurked mystery in the nature of biological system (plants and animals) will continue to be revealed through scientific evaluations. I recommend the book particularly to the high school students, undergraduate students and young scientists that are interested in the fascinating world of medicinal plant research.

**Professor J.K. Zjawiony**
School of Pharmacy
Department of BioMolecular Sciences
University of Mississippi,
Oxford, Mississippi 22 September, 2016

# Preface

Medicinal plant research is important for documentation, sustainable and effective utilization of the therapeutic potentials of these plants towards the management of human health challenges. The first edition of *Psychoactive Plants: Ethical Issues and Basic Evaluations* is a compilation of history, ethics in researches, botany, medical ethnobotany, pharmacological and phytochemical evaluations of medicinal plants. Traditional and preclinical knowledge of plants with central and peripheral nervous system effects are explored in the current edition. The challenges of beginners on why, where, what and how to proceed with their researches are effectively reflected upon. New investigators often lack sense of direction, history and bewildered in their pursuits of medicinal plant researches; hence, this book giving a captivating insight into botanical field works, herbarium, socio-cultural and environmental plant knowledge, laboratory and safety rules, animal care and use ethics, experimental models, plant description, biological activities, mechanism of actions, plants lists with their respective indications, reporting of experimental data and general research guides. Preclinical approaches to drug discovery and development are constantly undergoing conceptual shift and this has led to the use of numerous techniques or models. Although very few techniques are cited in this book for the illustration of synthetic and natural products screening, we acknowledge the existence of innumerous preclinical testing approaches. Readers are encouraged to consult the references provided for additional informations on issues of interests. Where possible, illustrative figures and chemical structures are included to facilitate comprehension of the book. Theoretical and practical knowledge, expertise and professional judgment are shared in this book to spur prospective young researcher and promote good science.

Finally, we owe a debt of gratitude to the editors of Astral International (P) Ltd. who patiently encouraged and guided us all through.

*James O. Fajemiroye*

# Contents

# Chapter 1

# Psychoactive Plants

☆ *James O. Fajemiroye and Elson A. Costa*

## ABSTRACT

*Recreational, social, religious or ritualistic importance of psychoactive plants that are literally considered as mind-altering plants have been widely reported over the years. The mind altering attribute of these plants could be associated with some of the expressions such as "beautiful head rush", "ecstasy", "trance induction", "numbing", "craving", "divine essence", "joy plant", "social tonics or party pills", etc. that are associated with their usage. These plants have different classes of active principle that works by interfering with the release, uptake or metabolism of neurotransmitters in the central nervous system. Some of their effect could be classified as stimulants, depressants, or hallucinogens. Some of which are addictive by leading to tolerance, psychological or physiological dependence. Among the classical examples of these plants include Papaver somniferum, Cannabis sativa, Erythroxylum coca, Nicotiana tabacum, Physostigma venenosum, Salvia divinorum, Atropa belladonna, Hypericum perforatum, Piper methysticum and Ephedra sinica. Other are Valeriana officinalis, Cola nitida, Catha edulis, Clibadium surinamense, Myristica fragrans, Rauvolfia serpentina, Datura metel, Theobroma cacao, Pausinystalia johimbe. This chapter has put together some historical background, basic chemistry and biological activities of psychoactive plants and their active principles.*

## 1.1 Introduction

Psychotropics, either from nature (plants and animals) or synthesized, are chemical substance that changes the way we feel, act or think. It is fascinating to observe behavioural changes in human or animal under the influence of psychotropics. Some of the feeling associated with the usage of these plants are sometimes expressed as "beautiful head rush", "ecstasy", "trance induction", "numbing", "craving", "divine essence", "joy plant", "social tonics or party pills", etc. These plants are capable of eliciting euphoria, dysphoria, heightened awareness, tranquility, stimulation, depression, convulsion, sedation, aphrodisiac, hypnosis, delirium, disorientation, amnesia, etc. Some are hallucinogen or narcotics that can change perception and cause vivid dreams. In addition to CNS properties of psychotropics, they sometime induce peripheral effects.

Some of the current knowledge and use of these plants are of serendipitous discovery. There are several historical instances with divergence report on the popular use of these plants. For instance, it was customary to give hemp seeds to the guests in order to promote hilarity and enjoyment (Galien, 1826); Calabar bean of *Physostigma venenosum* was used by calabar tribe in Nigeria as an "ordeal poison" in trials for witchcraft (Dworacek and Rupreht, 2002); nutmeg experience which was characterized by a drowsiness, a complete stupor and insensibility, delirious; the epic amazonian religious ceremonies involve a narcotic drink known as caapi in Brazil, yage in Colombia, and ayahuasca in Ecuador, Peru, and Bolivia from *Banisteria caapi, Banisteria inebrians, Banisteria quitensis, Banisteria rusbyana, and Tetrapteris methystica*; the roots of *Tabernanthe Iboga* are chewed by the natives to offset hunger and fatigue, and large doses produce excitement, mental confusion, and a drunken madness characterized by prophetic utterances (Tyler, 1966); the leaves of *Catha edulis* or Khat, are used daily by millions to invigorate the intellect and to assuage hunger. Hence, values (positive or negative) of these plants are traceable to culture, medicine, economy, abuse-recreation-addiction. The use of plants to influence brain function has long been essential to medical practice.

It is equally important to state that the influence of these plants on the peripheral nervous system could be associated with their medicinal application in different preparations (infusion, concoction, teas, tinctures, decoction, etc) and administration (inhaled as a snuff, swallowed, chewed, dermal application, bath, smoked) for the management of different medical conditions (abdominal irritation, abscesses, anaemia, arthritis, asthenia, chicken pox, conjunctivitis, constipation, cough, dermatitis, diabetes, diarrhoea, dysentery, fever, gastritis, gonorrhea, headache, hemorrhage, hypertension, itchy rashes, jaundice, leprous, malaria, measles, obesity, oedema, paralysis, poisoning, rheumatism, snake bites, sterility, syphilis, tapeworm, trypanosomiasis, urinary retention, vomiting, wounds, etc).

In the eighteenth century, Linnaeus classified certain plants that produced conscience and behaviour alterations under the generic name "Inebriantia" (Lewin, 1924). A century ago, Lewin proposed the first detailed classification to design the different pharmacological actions on the mind. He proposed a new generic groups: narcotic substances, and divided it in "euphorica" or mental sedatives, "phantastica" or hallucinogenic substances, "inebriantia", now limited to alcohol and some anaesthetics, "hypnotica" and "excitantia" . Linnaeus (1707-1788) provided a brief description and classification of the species (Linnaeus, 1742).

The struggle for survival and search for healing are key to ancient man resort to nature for remedy. The selection of plant for the treatment of specific diseases was based on the experiences gathered through "try and error" (Biljana, 2012). The 12 recipes for drug preparation is acclaimed to be the oldest written evidence of medicinal plants' usage for preparation of drugs (Kelly, 2009). The Ebers Papyrus, written circa 1550 BC, represents a collection of 800 proscriptions referring to 700 plant species and drugs used for therapy (Glesinger, 1954; Tucakov, 1964). According to data from the Bible and the holy Jewish book, the Talmud, during various rituals accompanying a treatment, aromatic plants were utilized such as myrtle and incense (Dimitrova, 1999). Over 2,500 years ago man first used a drug obtained from white willow bark, which was aspirin or acetylsalicylic acid. Today's scientists continue to be bewildered by just what aspirin's mechanisms of action are. Whatever the science of aspirin, an intelligent person today takes it just as our ancestors did for millennia (Riddle, 2002). Throughout time, explanations continue to vary just as purpose of administration do as well. Early 19th century was a turning point in the knowledge and use of medicinal plants. The discovery, substantiation, and isolation of alkaloids from poppy (1806), ipecacuanha (1817), strychnos (1817), quinine (1820), pomegranate (1878), and other plants, then the isolation of glycosides, marked the beginning of scientific pharmacy. With the upgrading of the chemical methods, other active substances from medicinal plants were also discovered such as tannins, saponosides, vitamins, hormones, etc. (Dervendzi, 1992).

Systematic study of cultural knowledge, belief and perception of indigenous plants (ethnobotany) revealed the use of psychotropic plants and their isolates for different purposes. These plants are found in the wild or even someone's backyard. The seed of *Physostigma venenosum* Balf. (Physiostigmine – Figure 1.1A), mind-altering effects of from *Salvia divinorum* Epling and Játiva (Salvinorin A – Figure 1.1B), hallucinogenic and aphrodisiac properties of *Atropa belladonna* L. (Atropine – Figure 1.1C), *Canabis sativa* L. (Phenolic terpenoid tetrahydrocannabinol – Figure 1.1D), analgesic property of *Papaver somniferum* L. (Morphine, Heroine (semi-synthetic), Codeine, Papaverine – Figure 1.1E), antidepressant effect of *Hypericum perforatum* L. (Hypericin – Figure 1.1F), anxiolytic effect of *Piper methysticum* L.f. (Pipermethystine, Dihydrokavain - Figure 1.1G), euphoric effect of *Erythroxylum coca* Lam. (Cocaine - Figure 1.1H), CNS stimulation by *Ephedra sinica* Stapf (Ephedrine - Figure 1.1I) and *Strychnos nux-vomica* L. (Strychnine – Figure 1.1J), *Galanthus nivalis* L. (Figure 1.1K) with Galanthamine (anti-AChE for Alzheimer treatment), sedative effect of *Valeriana officinalis* L. (Valerenic acid - Figure 1.1L), stimulant effect of *Cola nitida* (Vent.) Schott and Endl. (Caffeine – Figure 1.1M), stimulant and euphorigenic effects of *Catha edulis* (Vahl) Endl. (Cathinone – Figure 1.1N), convulsive property of *Clibadium surinamense* (Cunaniol – Figure 1.1O), mild sedative effect of *Nymphaea caerulea* Savigny (Aporphine - Figure 1.1P), stimulating property of *Nicotiana tabacum* (Nicotine – Figure 1.1Q), carminative property of *Myristica fragrans* Houtt. (Myristicin – Figure 1.1R), antipsychotics and sedatives effects of *Rauvolfia serpentina* (L.) Benth. ex Kurz (Reserpine – Figure 1.1S), epileptiform convulsions induced by *Anamirta cocculus* (L.) Wight and Arn. (Picrotoxinin - Figure 1.1T), calming effect of *Matricaria chamomilla* L. (Apigenin – Figure 1.1U), alleviation of pain with *Datura metel* L. (Scopolamine – Figure 1.1V), *Sceletium tortuosum* (L.) N.E. Br. as a mood elevator (Mesembrine – Figure 1.1W), neuroprotective and memory enhancing effects of *Ginkgo biloba* L. (Ginkgolide b - Figure 1.1X), stimulating property of *Theobroma cacao* L. (Theobromine – Figure 1.1Y), the stimulant and aphrodisiac properties of *Pausinystalia johimbe* (K.Schum.) Pierre ex Beille (Yohimbine – Figure 1.1Z) are indications of psychoactivities of these species and their respective active principles. Some of the active principles with psychoactive properties are alkaloids, glycosides, flavonoids, phenolics, saponins, terpenes, anthraquinones or steroids.

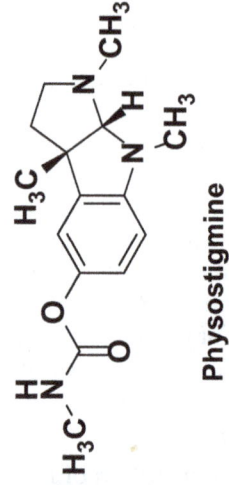

**Physostigmine**

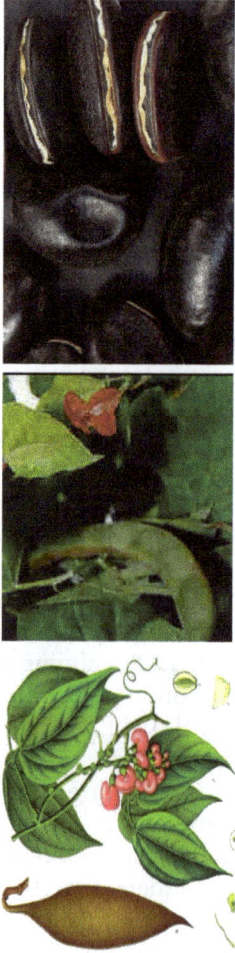

**Salvinorin A**

Figure 1.1A: *Physostigma venenosum* Balf..

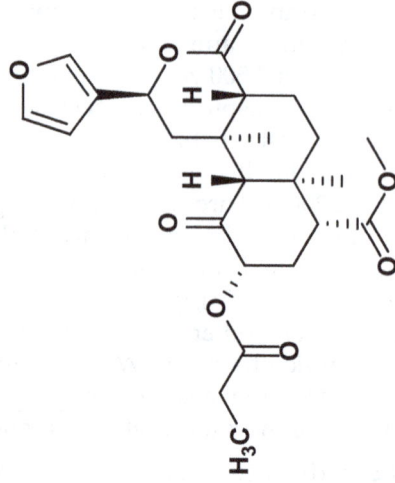

Figure 1.1B: *Salvia divinorum* Epling and Játiva.

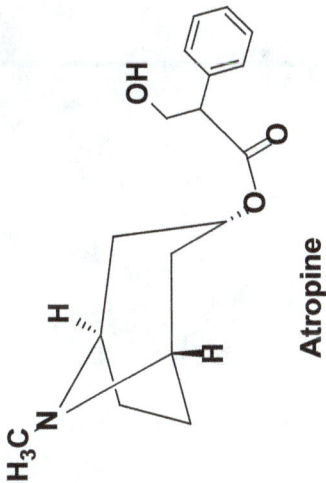

**Atropine**

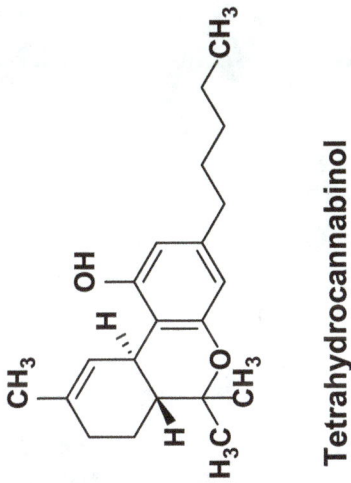

**Tetrahydrocannabinol**

**Figure 1.1C: *Atropa belladonna* L.**

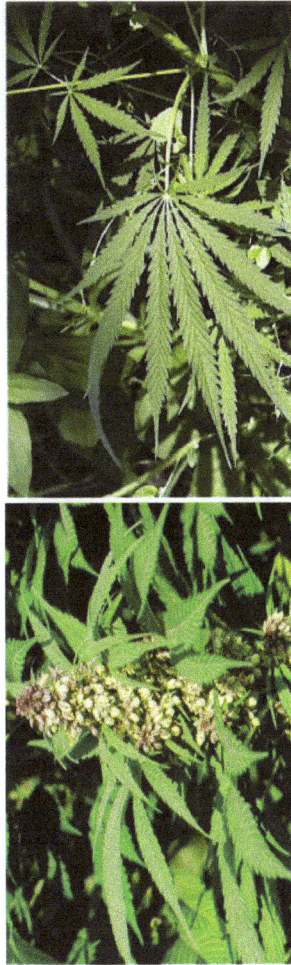

**Figure 1.1D: *Canabis sativa* L.**

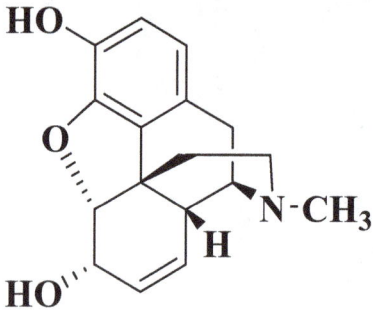

**Morphine**

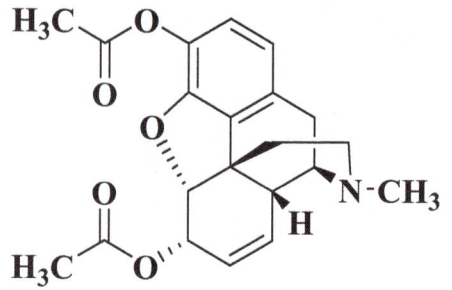

**Heroine**

**(semi-synthetic)**

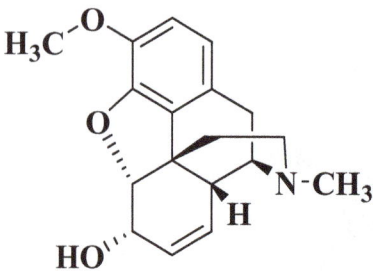

**Codeine**

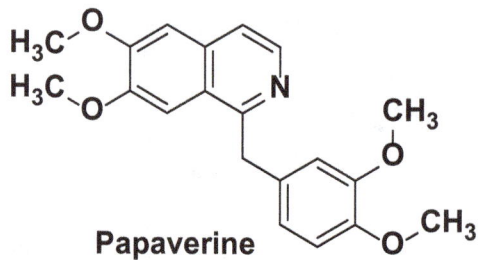

**Papaverine**

**Figure 1.1E:** *Papaver somniferum* L.

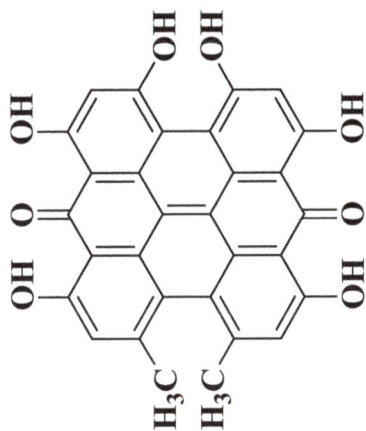

Hypericin

Dihydrokavain

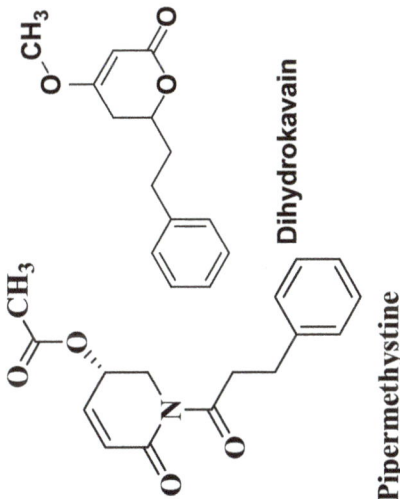

Pipermethystine

**Figure 1.1F: *Hypericum perforatum* L.**

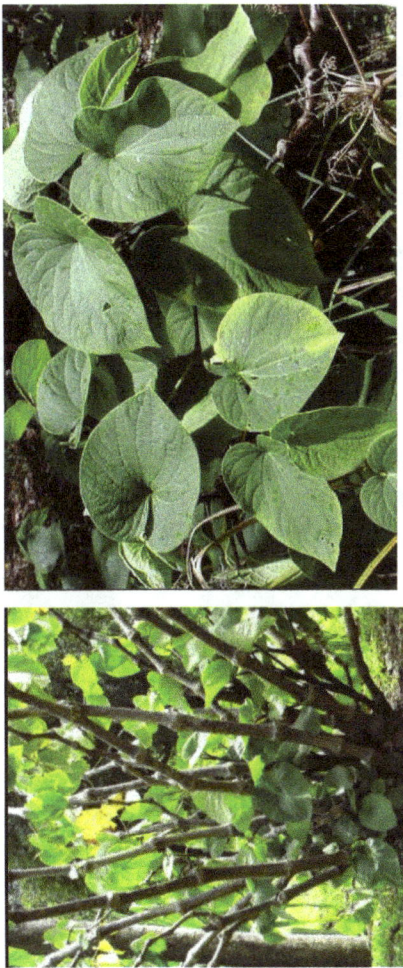

**Figure 1.1G: *Piper methysticum* L.f.**

Cocaine

Ephedrine

Strychnine

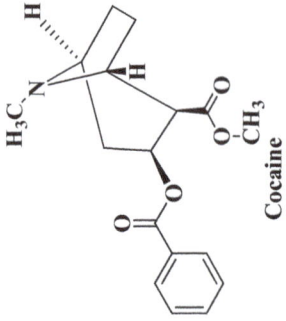

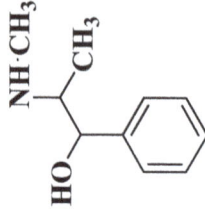

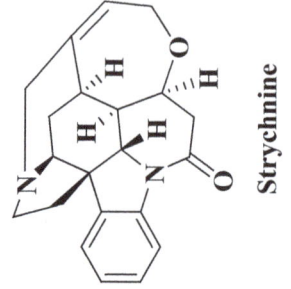

**Figure 1.1H:** *Erythroxylum coca Lam.*

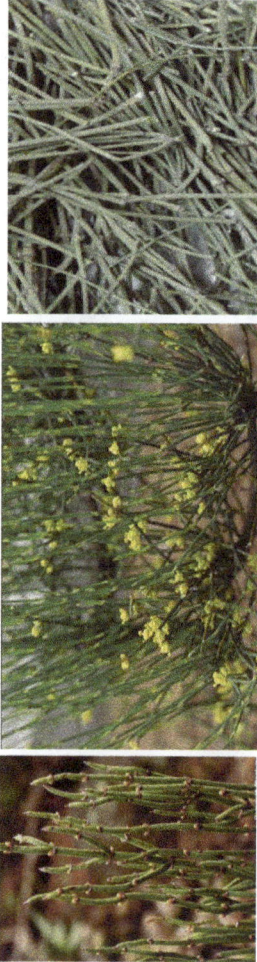

**Figure 1.1I:** *Ephedra sinica Stapf.*

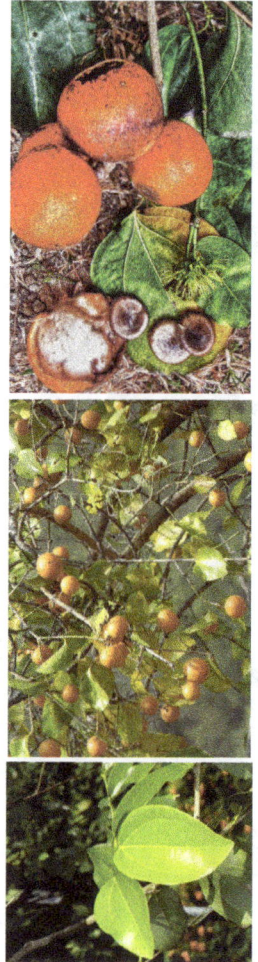

**Figure 1.1J:** *Strychnos nux-vomica L.*

Galanthamine

Valerenic acid

Caffeine

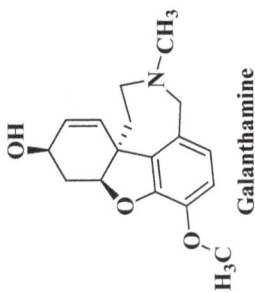

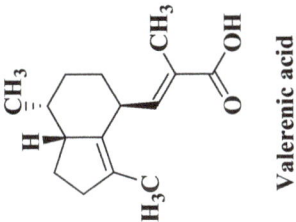

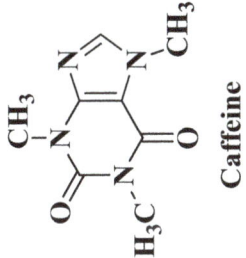

Figure 1.1K: *Galanthus nivalis* L.

Figure 1.1L: *Valeriana officinalis* L.

Figure 1.1M: *Cola nitida* (Vent.) Schott and Endl.

**Cathinone**

**Cunaniol**

**Aporphine**

Figure 1.1N: *Catha edulis* (Vahl) Endl.

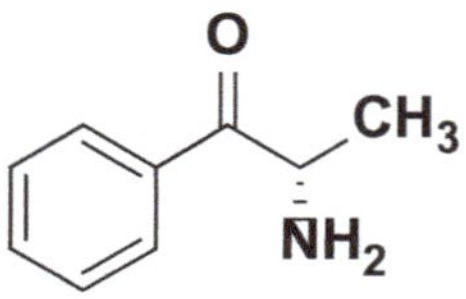

Figure 1.1O: *Clibadium surinamense.*

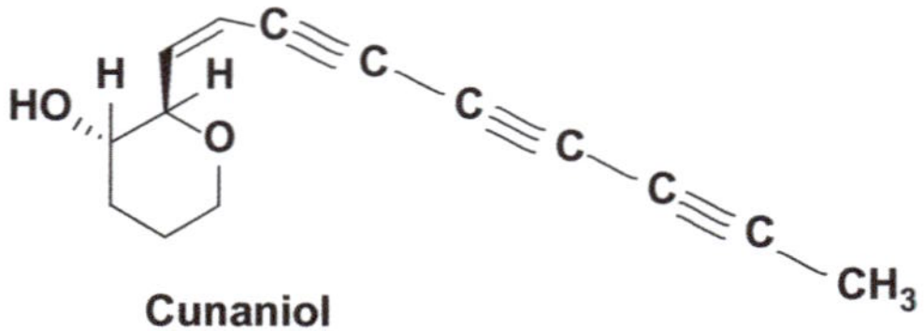

Figure 1.1P: *Nymphaea caerulea* Savigny.

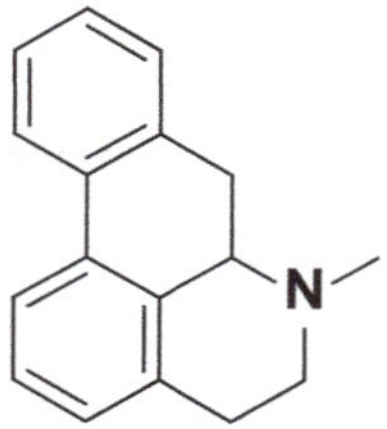

**Nicotine**

**Myristicin**

Reserpine

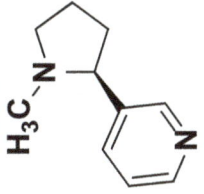

**Figure 1.1Q:** *Nicotiana tabacum.*

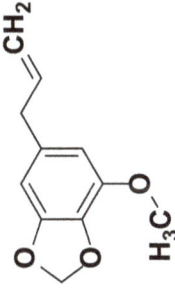
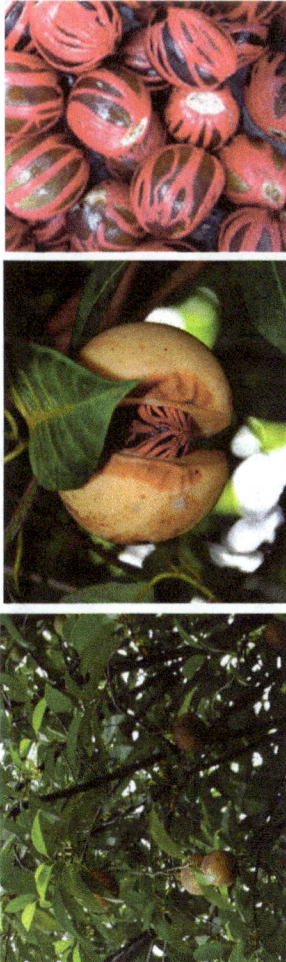

**Figure 1.1R:** *Myristica fragrans* Houtt.

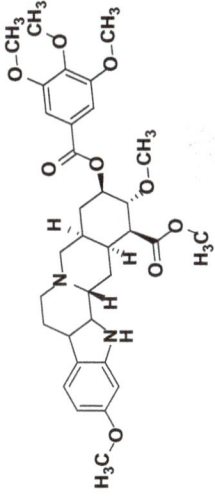
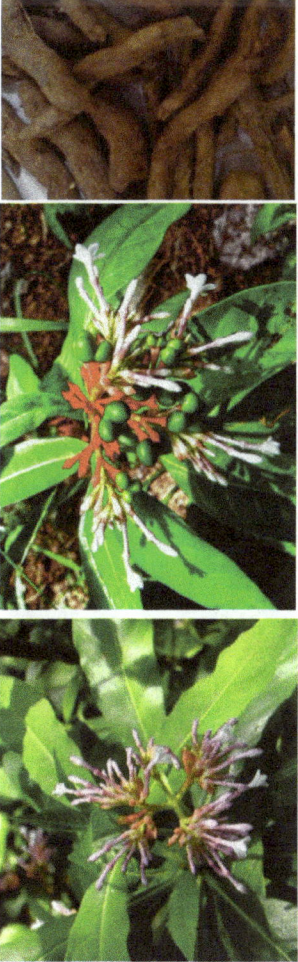

**Figure 1.1S:** *Rauvolfia serpentina* (L.) Benth. ex Kurz.

**Picrotoxinin**

**Apigenin**

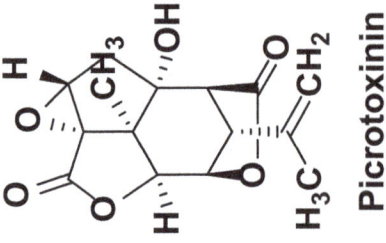

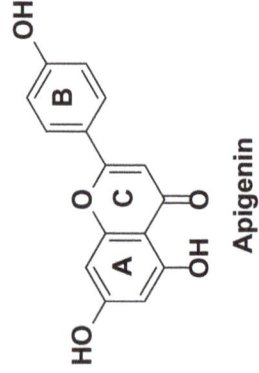

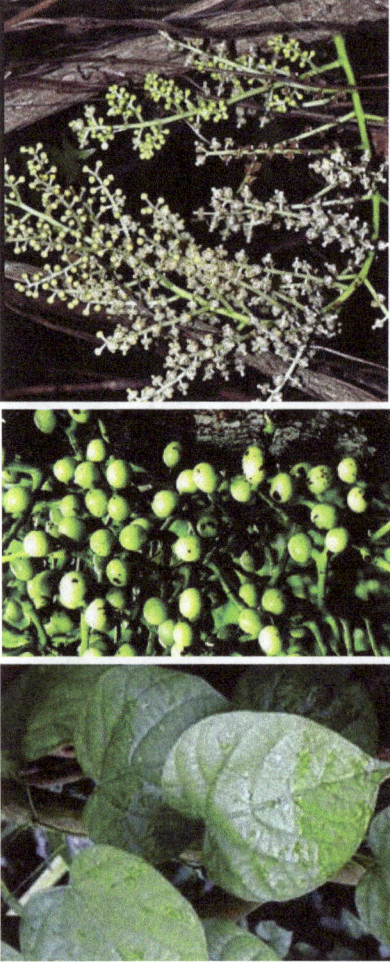

Figure 1.1T: *Anamirta cocculus* (L.) Wight and Arn.

Figure 1.1U: *Matricaria chamomilla* L.

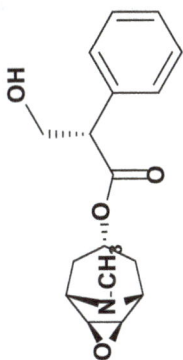

**Scopolamine Amnestic**

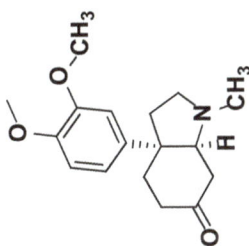

**Mesembrine**

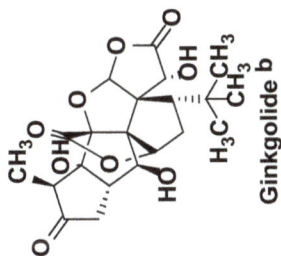

**Ginkgolide b**

**Figure 1.1V: *Datura metel* L.**

**Figure 1.1W: *Sceletium tortuosum* (L.) N.E. Br.**

**Figure 1.1X: *Ginkgo biloba* L.**

**Theobromine**

**Yohimbine**

Figure 1.1Y: *Theobroma cacao* L.

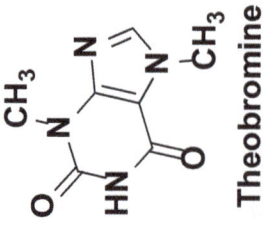

Figure 1.1Z: *Pausinystalia johimbe* (K.Schum.) Pierre ex Beille.

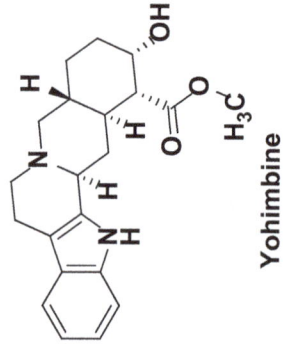

# 1.2 List of Some Psychoactive Plants

| | |
|---|---|
| **Acanthaceae** | *Barleria lupulina* Lindl.; *Ruellia geminiflora* Kunth. |
| **Adiantaceae** | *Adiantum capillus-veneris* L. |
| **Amaryllidaceae** | *Scadoxus puniceus* (L.) Friis and Nordal |
| **Annonaceae** | *Annona muricata* L.; *Annona senegalensis* Pers; *Cyathostemma micranthum* (A. DC.) J. Sinclair.; *Fissistigma lanuginosum* (Hook.f. and Thomson) Merr. |
| **Apiaceae** | *Arctopus echinatus* L. |
| **Apocynaceae** | *Calotropis gigantiea* (L.) R.Br. ex Schult; *Gomphocarpus physocarpus* E. Mey.; *Leptadenia reticulate* (Retz.) Wight and Arn. *Rauvolfia serpentina* (L.) Benth. ex Kurz; *Tabernaemontana pandacaqui.*; *Tabernaemontana sananho* Ruiz and Pav.; *Tabernanthe iboga* Baill.; *Xysmalobium undulatum* (L.) W. T. Aiton; |
| **Araceae** | *Pistia stratiotes* L. |
| **Asteraceae** | *Anacyclus pyrethrum* (L.) Lag.; *Artemisia afra* Jacq. ex Willd.; *Baccharis uncinella* DC.; *Bidens pilosa* L.; *Calea zachatechichi* Schltdl.; *Clibadium surinamense* L.; *Eclipta prostata* (L.) L. (Linn.); *Mikania cordifolia* (L. f.) Willd.; *Mikania scandens* (L.) Wild.; *Sphagneticola calendulacea* (Less) Pruski; *Tagetes lucida* Cav.; *Tagetes erecta* L.; |
| **Avicenniaceae** | *Avicennia officinalis* L. |
| **Aizoaceae** | *Sceletium expansum* L. Bolus; *Sceletium tortuosum* (L.) N. E. Br. |
| **Bixaceae** | *Bixa orellana* L. |
| **Cannabaceae -** | *Cannabis sativa* L. |
| **Cactaceae** | *Carnegiea gigantea* (Engelm.) Britt. and Rose; *Echinocereus triglochidiatus* Engelm.; *Lophophora williamsii* (Lem. ex Salm-Dyck) J.M. Coult. *Mammillaria senilis* Lodd. ex Salm-Dyck; *Pelecyphora asellifomis* Ehrenb. |
| **Campanulaceae** | *Lobelia tupa* L. |
| **Cupressaceae** | *Thuja occidentalis* L. |
| **Celastraceae** | *Catha edulis* (Vahl) Endl. |
| **Combretaceae** | *Terminalia glaucescens* Planch. ex Benth.; *Terminalia mollis* M. A. Lawson |
| **Convolvulaceae** | *Turbina corymbosa* (L.) Raf.; *Ipomoea violacea* L.; |
| **Coriariaceae** | *Coriaria ruscifolia* L.; |

| | |
|---|---|
| **Crassulaceae** | *Cotyledon orbiculata* L.; *Bryophyllum pinnatum* (Lam) Oken; |
| **Cyperaceae** | *Scirpus atrovirens* Willd.; Scleria hirtella Sw. |
| **Dioscoreaceae** | *Dioscorea dregeana* (Kunth.) T. Durand and Schinz |
| **Ericaceae** | *Pernettya furens* Klotzsch |
| **Erythroxylaceae** | *Erythroxylum coca* Lam. |
| **Euphorbiaceae** | *Croton sylvaticus* Hoschst.; *Jatropha gossypiifolia* L.; *Euphorbia hirta* L.; *Ricinus communis* L. |
| **Fabaceae** | *Acacia sieberiana* DC.; *Anadenanthera colubrina* var. *cebil* (Griseb.) Altschul; *Anadenanthera peregrina* (L.) Speg.; *Caesalpinia decapetala* (Roth) Alston; *Caesalpinia pulcherrima* (L.) SW.; *Daniellia oliveri* (Rolfe) Hutch. and Dalziel; *Clitoria ternatea* L.; *Detarium microcarpum* Guill. et Perr.; *Erythrina americana* Mill.; *Genista canariensis* L.; *Millettia grandis* (E. May.) Skeels; *Mimosa pudica* L.; *Mimosa tenuiflora* (Wild.) Poir.; *Mucuna pruriens* (L.) DC.; *Prosopis africana* (Guill. and Perr.)Taub.; *Rhynchosia phaseoloides* (SW.) DC.; *Senna singueana* (Delile) Lock; *Sophora secundiflora* (Ortega)DC.; *Tachigali paniculata* Aubl.; *Tetrapleura tetraptera* (Schum and Thonn) Taub.; *Trigonella foenum-graecum* L. |
| **Lamiaceae** | *Amasonia campestris* (Aubl.) Moldenke; *Clerodendron thomsoniae* Balf. F.; *Leucas longifolia* Benth.; *Melissa officinalis* L.; *Mentha aquatica* L.; *Mentha spicata* L.; *Salvia divinorum* Epl. and Jativa |
| **Lauraceae** | *Aniba canelilla* (Kunth) Mez; *Ocotea aciphylla* (Nees and Mart.) Mez |
| **Lecythidaceae** | *Couroupita guianensis* Aubl. |
| **Loganiaceae** | *Strychnos guianensis* (Aubl.) Mart.; *Strychnos nux-vomica* L. |
| **Lythraceae** | *Helmia salicifolia* (Kunth) Link.; *Trapa natans* var. *bispinosa* (Roxb.) Makino |
| **Malpighiaceae** | *Banisteriopsis caapi* (Spruce ex Griseb.)Morton; *Diplopterys longialata* (Nied.) W.R. Anderson and C.; *Tetrapteris methystica* R.E. Schult. |
| **Malvaceae** | *Urena lobata* L. |
| **Melastomataceae** | *Tococa guianensis* Mart. |
| **Meliaceae** | *Khaya senegalensis* (Desv.) A. Juss.; *Trichilia emetica* Vahl; *Trichilia pallida* SW.; *Trichilia quadrijuga* (Miq.) Kunth.; *Xylocarpus moluccensis* Lam. M. Roem. |

| | |
|---|---|
| **Menispermaceae** | *Abuta grandifolia* (Mart.) Sandwith; *Chondrodendron tomentosum* Ruiz and Pav. ; *Odontocarya tripetala* Diels |
| **Moraceae** | *Brosimum acutifolium* Huber; *Ficus anthelmintica* Mart.; *Helicostylis tomentosa* (Poepp. and Endl.) J. F. Mcbr.; *Maquira calophylla* (Poepp. and Endl.) C.C. Berg |
| **Myristicaceae** | *Virola calophylla* (Spruce) Warb.; *Virola calophylloidea* Markgr. ; *Virola elongata* (Benth.) Warb. |
| **Myrtaceae** | *Pimenta pseudocaryophyllus* (Gomes) Landrum |
| **Nyctaginaceae** | *Boerhavia coccinea* Mill. |
| **Nymphaeaceae** | *Nymphaea ampla* (Salisb.) DC. |
| **Orchidaceae** | *Oncidium cebolleta* Jaq. |
| **Ochnaceae** | *Ouratea castaneifolia* (DC.) Engl. |
| **Papaveraceae** | *Argemone mexicana* L.; *Fumaria indica* (Hausskn) Pugsley |
| **Phylanthaceae** | *Hymenocardia acida* Tul |
| **Passifloraceae** | *Passiflora edulis* Sims.; *Passiflora incarnata* L. |
| **Poaceae** | *Cymbopogon citratus* (DC.) Stapf; *Piresia leptophylla* Soderstr. |
| **Portulacaceae** | *Portulaca oleracea* L. ; *Portulaca quadrifida* L. |
| **Primulaceae** | *Cybianthus subspicatus* Benth. ex Miq. |
| **Proteaceae** | *Roupala obtusata* Klotzsch |
| **Phylolaccaceae** | *Phytolacca octandra* L. |
| **Piperaceae** | *Piper capense* L.f.; *Peperomia pellucida* (L.) Kunth |
| **Polygalaceae** | *Securidaca longepedunculata* Fres. |
| **Ranunculaceae** | *Nigella sativa* L. |
| **Rubiaceae** | *Bertiera guianensis* Aubl.; *Psychotria viridis* Ruiz and Pav.; *Randia armata* (Sw.) DC.; *Uncaria rhynchophylla* (Miq.) Miq ex Havil. |
| **Rutaceae** | *Pilocarpus pennatifolius* Lem.; *Spiranthera odoratissima* A. St.-Hil.; *Ruta chalepensis* L.; *Citrus sinensis* (L.) Osbeck; |
| **Salicaceae** | *Flacourtia indica* (Burm.f.) Merr. |
| **Santalaceae** | *Viscum album* L. |
| **Sapindaceae -** | *Erioglossum rubiginosum* (Roxb.) Blume; *Litchi chinensis* Sonn.; *Nephelium lappaceum* L.; *Paullinia cupana* Kunth; *Sapindus saponaria* L.; *Ungnadia speciosa* Endl. |

| | |
|---|---|
| Sapotaceae | *Mimusops elengi* L.; *Vitellaria paradoxa* C.F. Gaertn |
| Solanaceae | *Brugmansia arborea* (L.) Steud; *Brugmansia pittieri* (Staff); *Brugmansia* x *insignis* (Barb. Rodr.) Lockwood ex R.E. Schult.; *Cestrum laevigatum* Schltdl.; *Datura ceratocaula* Ort.; *Datura inoxia* Mill.; *Datura metel* L.; *Datura ferox* L.; *Datura stramonium* L.; *Petunia violacea* Lindl.; *Solandra brevicalyx* Standl.; *Solanum americanum* Mill. |
| Theaceae | *Camellia sinensis* (L.) Kuntze |
| erbenaceae | *Lippia alba* (Mill.) N.E.Br.; *Phylia nodiflora* (L.) Greene |
| Violaceae | *Rinorea guianensis* Aubl. |
| Vitaceae | *Cissus quadrangularis* L. |
| Zingiberaceae | *Hedychium coronarium* J. König |
| Zygophyllaceae | *Balanites roxburghii* planch. |

## 1.3 Some Centrally Mediated Biological Effects and Therapeutic Relevance

The hallucinogenic plants contain chemical compounds capable of inducing altered visual, auditory, tactile, olfactory, and gustatory perceptions or psychoses. The inexplicable mind-altering effects could be associated with its sacred important role in the religious rites and ancient traditions. It remained a means by which archaic men free himself from the prosaic confines of this earthly existence and to enable him to enter temporarily the fascinating worlds of indescribably ethereal wonder and communicate with supernatural realms. Only special people like kings and high priests were allowed to use these plants to maintain a constant intercourse with the spiritual world and possess powers that make them similar only to gods. These substances became the vehicle to preserve the natural order. Hallucinogenic plants are strange, mystical, and confounding. Although full potential of hallucinogen as aids to human needs is yet to be fully unraveled, this property could lead to the discovery of new drugs for the treatment of psychiatric illness. Examples of plants with hallucinogenic effect include; *Salvia divinorum* (also known as Sage of the Diviners, Ska María Pastora, Seer's Sage, and just Salvia). Its native habitat is in cloud forest in the isolated Sierra Mazateca of Oaxaca, Mexico.

Although the hallucinogenic properties of these plants are yet to be medically explored, the calming, sedative or tranquillizing property of some of these species have been explored in the treatment of psychiatric disorders. Some of these disorders are characterized by cognitive, somatic, emotional, and behavioural alterations with such symptoms as high blood pressure, elevated heart rate, sweating, fatigue, unpleasant feeling, tension, irritability, restlessness, low, sad or depressed mood, anorexia and anhedonia, among others. These symptoms constitute negative impact to the patient, families and society. The complexity of biological system and plural interaction of psychotropic plants or isolates still keep a lot of neuropharmacologist groping. As this psychoactives are capable of crossing blood brain barrier and sufficiently available in different brain targets/areas (hippocampus, amygdala, hypothalamus, prefrontal cortex, ventral tegmental area, locus ceurulus, locus coeruleus, raphe nucleus, paraventricular nucleus, etc), it is unsuprissing that they have plural mechanisms of action which confer therapeutic application in the treatment of several mental illnesses such as depression, Alzheimer, Parkinson, schizophrenia, epilepsy and insomnia among others.

## 1.4 Mechanism of Psychoactivities

The understanding the many mechanisms by which psychoactive plants or its isolates influence brain function is critical to predicting the biological activity and safe use. According to Lewin (1924) and Nichols (2004), detailed studies investigating the molecular mechanisms of psychoactivities could yield clues to the "chemistry of consciousness". Psychoactive drugs could induce biological activity by mediating the synthesis, transport, release, biotransformation (anabolism or catabolism) of neurotransmitters [acetylcholine (ACh), γ-aminobutyric acid (GABA), glycine, serotonin (5-HT), dopamine (DA), norepinephrine (NE) – Figures 1.2 A, B, C, D respectively] like some synthetic compounds [fluoxetine and citalopram (serotonin reuptake inhibitor – Figures 1.2E and F, respectively), imipramine (monoamine reuptake blocker - Figure 1.2F), iproniazid and tranylcypromine (monoamine oxidase blocker - Figures 1.2G and H, respectively)] or mimicking the actions of neurotransmitters on different biological systems (serotoninergic, GABAergic, glutamatergic, glycinergic, catecholaminergic (dopaminergic and noradrenergic), endocanabinoid, peptidergic, hypothalamic-pituitary-adrenal pathways, etc).

The mechanisms of psychoactivities could be traced to agonism or antagonism of opioid, cannabinoid, CRF, vasopressin, glucocorticoids, NK, NMDA, melatonin, galanin, NPY receptors among others. Cytokine-regulated pathways, enzyme inhibition or activation (histone deacetylase, AChE, phosphodiesterase, adenil cyclase, cyclooxygenase, tyrosine hydroxylase, tryptophan hydroxylase, monoamine oxidase, etc), tissue plasminogen activator, neurotrophic factors related mechanisms, blockade of voltage-gated sodium ion channel, potentiation of gamma-aminobutyric acid binding to GABAA receptors and opening or blockade of ion channel.

Mesembrine (Figure 1.1W), a serotonin-like alkaloid and the active constituent of *Sceletium tortuosum* (a species that have been used by South African shamans from prehistoric times as "mood enhancer"), is a potent serotonin re-uptake inhibitor. This mechanism suggests its potential for the treatment of anxiety and depression. Interaction of this compound with nicotinic, dopamine and nor-adrenaline receptors could be attributed to the traditional mood-elevating uses, and suggest additional therapeutic potential. The β-carboline derivatives such as harmine and harmaline with indolamine-like chemical structure sometimes acts as a 5HT receptor agonist to induce hallucinogenic effect. Similarly, *Annona muricata* L. which is acclaimed for its anxiolytic property has isoquinoline alkaloids (anonaine, nornuciferine and asimilobine) block 5HT1A receptors (Hasrat *et al.*, 1997). Liriodenine that is found in *Cyathostemma argenteum* (Khamis *et al.*, 2004) is known to block muscarinic receptors. Ibogaine (an isolate of *Tabernanthe iboga*) which was reported to reduce

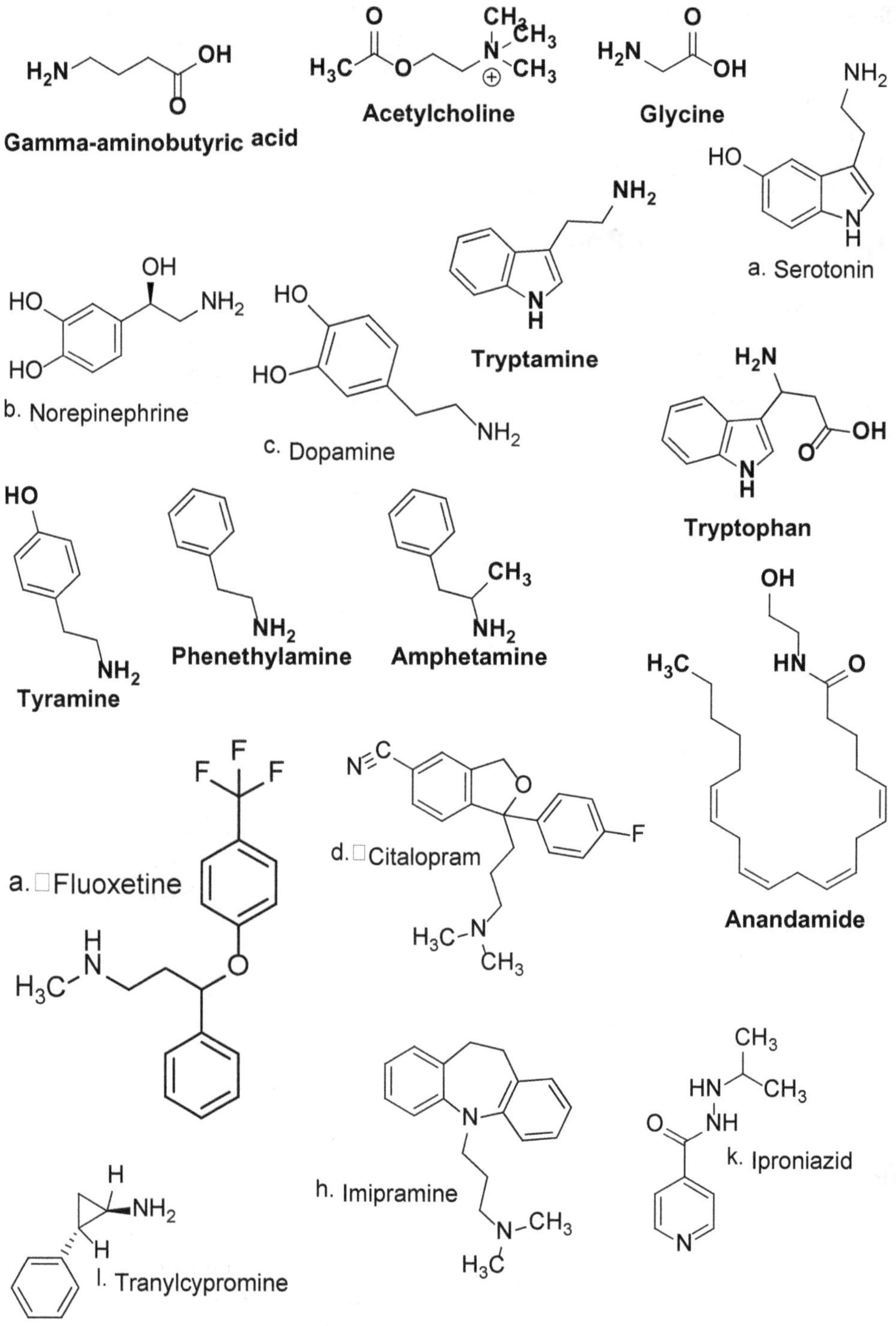

**Figure 1.2: Chemical Structure of some Neurotransmitters, CNS Acting Phytoconstituents and Synthetic Compounds.**

drug craving has a mechanism associated with regulation of the serotoninergic system that, in turn, regulates dopamine release (French *et al.*, 1996). Potential therapeutic application of this compound for the treatment of Alzheimer's disease and Huntington's chorea, among others, has been associated with its interaction with N-methyl–D-aspartate neuron receptors against excessive release of excitatory amino acids. Psychollatine, an indole alkaloid isolated from *Psychotria umbellate*, is centrally active via serotoninergic 5-HT2 (A/C) receptors (Both *et al.*, 2004). 5-HT2A receptors have been identified as the principal target for hallucinogen actions (Glennon *et al.*, 1984).

Some of the psychotropic plants or active principles could elicit allosteric modulation of the $GABA_A$ receptor complex. Unlike benzodiazepine (diazepam), picrotoxinin (a sesquiterpene, which is found notably in the seeds *Anamirta paniculata*) is a noncompetitive antagonist of $GABA_A$ that impede the GABAergic presynaptic inhibition of excitatory transmission of primary afferent neurones of the spinal cord, hence a general increase in neuronal activity, alertness, anxiety, spasms, seizures, and even death. Series of $\alpha$-pyrone including kawaine (isolate of *Piper methysticum*) elicits anxiolytic effects through GABAA receptor binding (Klohs *et al.*, 1959; Jussofie *et al.*, 1994). The anxiolytic and sedative effects of *Valeriana officinalis* involve the GABAergic system. The displacement of [$^3$H]muscimol from GABAA receptor by crude extracts of *Valeriana officinalis* L., the inhibition of the firing rate in most brainstem neurons by the extract of this species and valerenic acid and subsequent blockade of their effect by bicuculine further support gabaergic mechanism. The 6-methylapigenin from rhizomes and roots of *Valeriana wallichii* has been reported as a competitive ligand for GABAA receptor (Wasowski *et al.*, 2002). The calming property of *Matricaria chamomilla* could be associated with the presence of Apigenin, or 5,7,4 -trihydroxyflavone. It has been reported that this flavone blocks flunitrazepam binding to GABAA receptors, displaces flumazenil from the central benzodiazepine binding site, and reduces GABAactivated chloride channels. Hypericum perforatum is one of the most widely used psychoactive plants with putative antidepressant actions (Fajemiroye *et al.*, 2016). Purified substances obtained from *H. perforatum* can interact with a variety of biogenic amine and peptide receptors with low affinities, generally in the micromolar range (Roth *et al.*, 2004). Butterweck *et al.* (2002) reported that amentoflavone has a high affinity for the GABAbenzodiazepine receptor complex and y-opioid receptors while hypericin has moderate affinity for CRF-1 receptors.

Glycinergic system is another pathway with inhibitory neurotransmitter glycine which can be found on moto-neurons that operates via the activation of distinct postsynaptic receptors. The mechanism of strychnine (an indole alkaloid present in the seeds of Strychnos nuxvomica) action has been associated to glycinergic antagonism.

The $\mu$-opioid receptors have been reported as the main site of analgesic action of morphine, (Pert and Snyder, 1973). Salvinorin-A isolated from Salvia divinorum (a hallucinogenic plant) was discovered that to be a potent and selective $_o$-opioid receptor agonist (Roth *et al.*, 2002; Chavkin *et al.*, 2004).

Ephedra extracts contain a complex mixture of phenylpropanolamines (Rothman *et al.*, 2003) with several isomers of ephedrine. Ephedrine and related phenylpropanolamines was presumed to act directly on postsynaptic $\alpha 1$-adrenergic receptors (Gilman *et al.*, 1992). Subsequent works of Roth *et al.* (2004) showed that ephedrine and other phenylpropanol amines, had their highest affinities as norepinephrine transporter substrates with affinities in the 10- to 40-nM range. The main cardiovascular actions of ephedrine and related phenylpropanolamines are due to an indirect sympathomimetic action and not to direct activation of postsynaptic adrenergic receptors (Roth *et al.*, 2004).

The blockade of adenosine receptors by caffeine increases the activity of dopamine, which is implicated in the effects of caffeine (Cauli and Morelli, 2005). Like amphetamine (Figure 1.2), increase in the levels of dopamine in the brain has been implicated in the CNS stimulant effect of cathinone (isolate of *Catha edulis* – Figure 1.1) (Patel, 2000; Connor *et al.*, 2002). Cathinone had high affinities as dopamine transporter substrates with affinities in the 14- to 18-nM range (Roth *et al.*, 2004). It is likely that the high affinity of cathinone for dopamine transporters is important for their high abuse potential. Ephedrine and other phenylpropanol amines have modest activity as dopamine transporter substrates with affinities in the 200- to 2000-nM range.

The ginkgolides B 20-carbon cage molecules with six 5-membered rings, is an antagonist of platelet-activating factor (PAF) that has numerous biological effects (Koltai *et al.*, 1991). Platelet-activating factor has a direct effect on neuronal function and long-term potentiation (Del Cerro *et al.*, 1990; Wieraszko *et al.*, 1993). Flavonoids of ginkgo extract contribute to ginkgo's antioxidant and free radical scavenger effects (Oyama *et al.*, 1994). These effects protect brain neurons against oxidative stress induced by peroxidation (Maitra *et al.*, 1995; Ni *et al.*, 1996); *G. biloba* extract has been reported to inhibit monoamine oxidase A and B. This effect was associated to the presence of kaempferol (Sloley *et al.*, 2000).

Reports have shown the action of active principle $\Delta$-9-tetrahydrocannabinol (THC), a phenolic substance in *Cannabis sativa*, on anandamide receptor (CB1). The caffeine and theobromine (purinic alkaloids), 2-Phenylethylamine, tryptamine and tyramine (amines - Figures 1.2J respectively) are found in cocoa beans of *Theobroma cacao*. Like caffeine, theobromine is a stimulant substance. The effects of this compound on the CNS include increased concentration and attention. Theobromine is a competitive nonselective phosphodiesterase inhibitor (Essayan, 2001), which raises intracellular cAMP, activates PKA, inhibits TNF-alpha (Deree *et al.*, 2008; Marques *et al.*, 1999), leukotriene synthesis, inflammation and innate immunity (Peters-Golden *et al.*, 2009). Substances found in chocolate (commercial product from cocoa seeds), such as phenylethylamine, analog of amphetamine (Figures 1.2 E and F, which is also present in the brain), theobromine (Figure 1.1Y), anandamide (Figure 1.2G) and tryptophan (Figure 1.2H) trigger mood enhancing/euphoric compounds and neurotransmitters to be released in the brain. Theobromine is considered as the compound that makes chocolate special. Phenylethylamine is believed to cause mesolimbic release of dopamine. This may be why women prefer chocolate to sex as the thresold to realese this pleasurable catecholamine or attain its peak level without orgasm (unconfirmed speculation). However, tests have shown that most of the phenylethylamine in chocolate is broken down before it reaches the brain. The mechanisms of chocolate cravings is still unclear. Anandamide could mimic the effects of THC from marijuana. Low level of anandamide in chocolate and its break down by stomach acid makes it impossible to elicit pleasurable "high" like marijuana (unconfirmed).

## Chapter 2

# Ethnobotanical Approach to Psychoactive Pre-Colombian Plants from Ancient and Contemporary American Cultures in Mexico and Chile

☆ *Roberto Saavedra*

## ABSTRACT

*This chapter has the intention to review various aspects of ancient and contemporary use of hallucinogens among peoples from Mesoamerica in Mexico, in comparison with others used in South America, particularly in Chile. Another objective is to see the existence of parallelisms and differences among hallucinogenic plants usage from indigenous groups of both places. From prehistoric ages, it is known that the knowledge of plants began with the use of them by the first human groups on earth. They started to search for plants from their surrounding areas and used them primarily for food and medicine. Along with those first plants knowledge, they used psychoactive plants for specific purposes such as protection, contact with the spirits, hunting, and prayed for rain and food supply by nature. Even though, medicinal and psychoactive use of plants is still very active nowadays and has been transmitted by family heritage and upgraded through time, the tradition of plant use has been coming down from generation to generation unfortunately. This plant wisdom and knowledge have been flowing in two parallel streams, culture-lore and folklore, which both are important for its subsistence. In this sense, ethnobotanical studies are needed to perform in order to collect and understand the traditional knowledge with more detail about its cultural and practical meanings for the psychoactive plant usage in order to maintain these practices alive into the indigenous context and how this set of knowledge can be used to open new fields of exploration and research in clinical and pharmacological studies in modern societies.*

## 2.1 Introduction

Many plants are toxic as its etymology explains the meaning of them. Toxic in Greek, *toxikón*, means 'poison arrow.' This concept emphasizes that the medicinal plants can cure ailments by being toxic or poisonous. Hallucinogenic plants should be considered toxic ones. They produce intoxications or states of trance (Schultes and Hofmann, 2012).

Psychotropic plants have been present in many cultures at different times and they are recently subjected to a constant reinterpretation of their uses and purposes. They have been called magical or vehicles to communicate with gods, ecstasy tools that modify perceptions in human minds of the reality and cause states of confusion. These plants are those that guide the soul (from the Greek *psyke*, 'soul' and *tropos*, 'to guide', 'direct'). Given that a colloquial term like *soul* still conserves its mysterious deep rootedness, while its definition depends on specific cultures and the knowledge of each individual, we are obliged to resort to myth, legend and the fragile history of differing peoples for an explanation (Lozoya, 2003).

It is believed that probably the usage of wild flora and first humans that inhabited the American continent were developed in a parallel way during time. In the case of Mexico, plant resources were used as food, medicine, and daily life elements. That was the reason why these plants were considered as sacred for divinatory purposes, such as the use of mushroom by indigenous groups. Some of the most important species cited by Álvarez (2003) were *ololiuhqui* (*Turbina corymbosa*), *péyotl* (*Lophophora williamsii*), *tlitliltzin* (*Ipomea violacea*), and *teonanácatl* (*Psilocybe aztecorum*).

Most of these prehispanic cultural groups in Mesoamerica and their health systems were, and still are, mainly dependent of herbalism. Among nahuas, mayans, and other important mexican groups, health was based on the equilibrium and combination of natural, corporal, and supernatural forces, where medicinal plants played a significant role by their interaction among them. Typical practices for prevention and cure of illnesses were applied by shamans over particular areas of the body either to individuals as well as whole groups. These parts were located in the strong and weak animic centres besides vital forces (Bye, 1999).

For aztecs, there were three main animic centres. The first one was on the head, *tonalli*; second on the heart, *teyolia*, and the third on the liver, *ihíyotl* (Bye, 1999). They also knew that, for some illnesses, spiritual world intervention was needed to find the causes and remedies. For that reason, plants with the capacity to transform and altered the perception were used to conduct users to other worlds with their minds and gather information on how to know the reasons and plant remedies by this mystical approach. That is why shamanic ceremonies were offered to predict the future and cure ailments of the body and soul. Thus, the term hallucination is considered an inner to human condition, so dreams, revelations, and visions are expressed themselves differently throughout the world.

Nonetheless, traditional medicine does not share the same medical principles as modern medicine, its empirical and scientific substrate is rational. An experimental study over 118 aztec medicinal plants identified in colonial documents showed

85 percent of them contained biochemical compounds that could have desirable effects. By comparing the previous statement with modern medicine, it can be concluded that 60 percent of those medicinal plants can be considered effective (Ortiz de Montellano, 1990). At the same time, most of the plants used as entheogens contained alkaloids that have an effect on the central nervous system altering senses and sensitizing them to external stimuli (Schultes and Hofmann, 1980).

As a supportive argument, direct archeological evidence also exists for medicinal plants. Cultural fragments were found to be part of the evidence showing many parts of the plants prepared for usage in a medical context (Montúfar López, 1985; Montúfar López, 1998) as well as archeoethnobotanical reports also add more multiple plant uses such as food, textile, wood aside from medicinal uses.

Other evidences associated with health practices are the indirect ones used by ancient shamans from Cave-Maker in Chihuahua, Mexico. They used reed (*Phragmites* sp.) for inhalation of pulverized plants as modern Tarahumaras still use nowdays. This shamanic practice is associated to diagnosis and treatment in divinatory ceremonies and rituals. In other parts of Mexico such as Colima, Guerrero, Nayarit, and Oaxaca different forms of pipes were made with decorative animal figures for snuffing (Furst, 1974). Iconography is also another indirect evidence to demonstrate the presence of medicinal remedies. Xochipilli, the prince of flowers, illustrates how aristocratic aztecs used entheogenic plants for curative rituals and divinatory practices. Several plants were hewed on *Xochipilli* statue such as tobacco flowers (*Nicotiana tabacum*), Ololiuhqui (*Turbina corymbosa*), sinicuichi (*Heimia salicifolia*), and cacao flower (*Quararibea funebris*).

Acculturation from Spaniards allowed the observation of transformations of plant usage and similarly the permanence of health practices until now, as several documents from colonial times such as Codex Badiano (*Libellus de Medicinalibus Indorum Herbis*) ratified it.

## 2.2 Plants and Visions

Entheogenic plants are considered vehicles that can cause, by ingestion, religious experiences to human beings. Ancient societies used psychodelic plants to transport their inner mind to other spheres of existence. This phenomenon is common to non-industrialized societies where divine plants are expected to be well known by shamans of tribes in order to transfer relevant information from the god plants to the mind. There is a communication between the shaman who consumes the plant and the different worlds perceived by himself.

The way this type of plants make an effect over the people's mind is through effectively creating a bridge between their mind and a superior world. In the case of a person that analyzes these experiences from outside of the context of a shaman, these personal experiences are conceived malign and full of drastic and horrifying conceptions inside of the mind. Lipp (2002) states that the effects are quite diverse and are dependent from the interaction between the environmental conditions of the place where is taking place, the ritual and the mental environment of the user's mind. This process has unique characteristics such as emotional states, cultural scenery and customs, own personality, future expectations, and motivations from the user.

People from this cultural context have a deep respect to entheogenic plants for the role they play into the religious practices taking place inside their culture. Exists codes carried out finely by the shaman to create specific environment with particular purposes such as purification of a person or places as well as diagnose of an illness.

A person who is not immerse into this kind of cultural aspects and is not knowledgeable of this type of ritualistic concepts and religious practices consider entheogenic plants as merely objects of pleasure and own satisfaction through consumption being part of illegal practices and punishment (Lipp, 2002).

## 2.3 Sacred Plants

Plant world is very powerful and hermetic to most of the common people where reality functions totally differently from what is expected. In this world, chosen people are shamans who possess knowledge and instructions from ancient times for the usage and protection of sacred plants. People from the American continent, specifically Mesoamerica and South America, have a profound knowledge and consider sacred plants as divinities, so shamans from these regions, when ingest psychoactive plants, have confessed that they listen in their inner body sounds and perceive images from plants which send special messages to them to cure people from their community and also carry other specific meanings. After having a long term of training into the forest and mastering of ecstasy, shamans acquire abilities to diagnose and heal people magically by the use of divinatory chants and plant gathering ritualistically.

More often, a long spectrum of areas of study, have been deeply interested in sacred plants, which have psychoactive properties as well. These plants have been subject of extensive research from a multidisciplinary standpoint in fields of study such as linguistics, history, mythology, ethnology, theology, biochemistry, botany, pharmacology, and physiology; all dealt with plants in someway or another to reinforce a combination of aspects that enhance the description and better detail documentation of them.

An important progress was obtained at the end of the XIX century in which research of active components from traditional plants was the primary goal. This was the isolation of morphine from opium (*Papaver somniferum*) and cocaine from coca plant (*Erythroxylon coca*) in 1817 and 1859, respectively. Later on in 1898, mescaline was isolated from peyote cactus (*Lophophora williamsii*), which exhibited psychological effects in humans by being considered an important visionary plant to indigenous groups of Northern Mexico. Hallucinations, intense emotional swings, thought pattern modification and ecstasy were some of the effects presented by ingestion of this cactus (Díaz, 2003).

Besides, an ethno-mycological study was also undertaken by Wasson and Roger Heim in 1958 with their published book so called Hallucinogenic mushrooms of Mexico, in which they described the ritual use of fungi of the *Panaeolus* genus by Maria Sabina, a zapotec shaman female.

## 2.4 Psychoactive Plant Species

The following classification of psychodysleptic species and their concepts associated with their characteristics are considered of special attention and interest by the correlations between psychological basis and ethnological, chemical and cerebral characteristics.

Hallucinogens are biological substances that produce hallucinations or perceptions without a confirmed object. Another definition said by NIH (2016) indicates that hallucinogens are a diverse group of drugs that alter perception (awareness of surrounding objects and conditions), thoughts, and feelings. They cause hallucinations, or sensations and images that seem real though they are not. Hallucinogens can be found in some plants and mushrooms. Some important sacred plants par excellence are examples of hallucinogens such as *peyote* (Figure 2.1) and mescaline, mushrooms and their psilocybin (Figure 2.2) and LSD, derived from ergot; *Yopo* (*Anadenanthera peregrine*) that contains N-Dimethyltryptamine which is used by Amazonian tribes as an inhalant. Besides, they share ritual encounters as

**Figure 2.1:** *Peyote.* **A plant for shamanic rituals in Mexico**
**@ Frank Vincentz.**

Figure 2.2: *Teonanácatl*
(*Psilocybe mexicana*).
Taken from Mycotek.org.

Figure 2.3: Hoja de la pastora or
*pipiltzintzintli* (*Salvia divinorum*).
@ Eric Hunt.

Figure 2.4: Ololiuhqui
(*Turbine corymbosa*).

Figure 2.5: Toloache
(*Datura stramonium*).
@ Roberto Saavedra.

seen with Huichol pilgrimage to Wiricuta or *veladas* with chants of María Sabina. These gatherings contain elements of a huge complexity when using these divinatory sacred plants (Díaz, 2003).

By looking at the molecular level, hallucinogenic molecules have an effect on neurons that produce serotonin (5-hydroxy-tryptamine), acting on nervous system by affecting the perception and emotions. Furthermore, dopamine is produced either by the brain or peyote changing to mescaline. Thus, the phenomenon of consciousness can be expressed by the human brain, and specifically, by the nervous system, which has an intimacy relationship with the environment and its ecological elements along with the social system allowing of them the presence of hallucinations.

Trance inducers are ancient plants that are used in rituals and divinatory environments. They can produce lethargy to people who consume them during this ceremonial processes. These people get stimulus to their memory and enhancement of personal perception that eventually can be irritating. Some examples of the sacred plants are *ololiuhqui* (Figure 2.4), or seeds of the Virgin (Morning Glory – *Ipomoea violacea* and *Rivea corymbosa*), which are usually and traditionally used by in southern people from Mexico, specifically in Oaxaca. At a molecular level, ergotamine alkaloids produce truly hallucinatory effects (Díaz, 2003).

Cognodysleptics are drugs that stimulate imagination enhancing sensations and fantasy without producing hallucinations; they just alter some memory mechanisms, making hard to recover information. The compounds responsible for this mechanism of action are terpenes present in marijuana (*Cannabis sativa*), which is the main representative plant of this segment. The ancient plant is originally from Asia, but was introduced to America during the colonial time. Afterwards, indigenous people used it for ritual purposes having an indigenous name in *Náhuatl*, *pipiltzintzintli*. Recently, it is known that the brain produces endogenous compounds similar to tetrahydrocannabinol (THC), acting on natural brain receptors. These endocannabinoids interact with specific neuronal nets in correspondence with endorphins, serotonin and dopamine. Their identification as lipid ligands has triggered and exponential growth of studies exploring the endocannabinoid system and its regulatory functions in health and disease (Sides, 2015; Pacher *et al.*, 2006). Similar effects are produced by *hoja de pastora* (*Salvia divinorum*) (Figure 2.3) and the Chontal *hoja madre* (*Calea zacatechichi*) from Mazatec rituals. They are used for dream interpretation and are able to modify sleep phases.

Deliriogens are the last group of plants having different effects from the latter ones. They reduce consciousness and may produce a delirium similar to fever when taken in high doses, causing disorientation and intense hallucinations that a person can confuse with the external reality. They are considered true narcotics. Some examples of them are *toloache* or *tlápatl* (*Datura stramonium*), wild tobacco or *yetl* (*Nicotina rustica*), plants that are denoted to have extremely potent effects. These plants of ancient times are used secretly and have a hidden and obscure tradition, generally for rites of witchcraft to harm enemies or confuse an unfaithful person. Their main active compounds are tropane alkaloids (*i.e.* scopolamine) that block acetylcholine-specific receptors found in *Solanaceae* genus. *Toloache* (Figure 2.5) in Mexico and henbane in Europe belong to this plant family (Díaz, 2003).

## 2.5 Archeological Evidences

The traditional use of psychoactive plants by Mexican indigenous peoples is ancient and very prolific, which is reflected in archeological ruins all over the Mexican territory highlighting the knowledge of states and mental effects. For instance, in Casas Grandes, an archeological site in northern Mexico, old ceramics were found with painted spirals to represent *peyote* (Figure 2.6). Nowadays, Huichol groups also use these spirals in their clothes decorations and artifacts for the same purpose (Figure 2.7). In 10 000 year-old tombs from southern Texas, another deliriogen plant, mescal bean (*Sophora secundiflora*), was found; this one was replaced by *peyote* later on. In Tula, Hidalgo, two reliefs portray a *Chac Mool* where *toloache* sprouts emerge from its central part of its body (Figure 2.8) representing the act of psychoactive plants usage and their importance for rituality in old times. In terms of sculptures, exist representations of mushrooms related to ceremonies for deities in archeological Maya sites as well. In codices such as Codex Florentino, psychotropic plants were mentioned by friar Bernardino de Sahagún with their indigenous names: *ololiuhqui, péyotl, tlápatl, tzintzintlápatl, míxitl,* and *nanácatl* (Díaz, 2003).

All these sacred places are very important representations of knowledge and remnants of splendor of ancient cultures that can give more insights in respect to the usage of psychoactive plants and how they can modify states of consciousness.

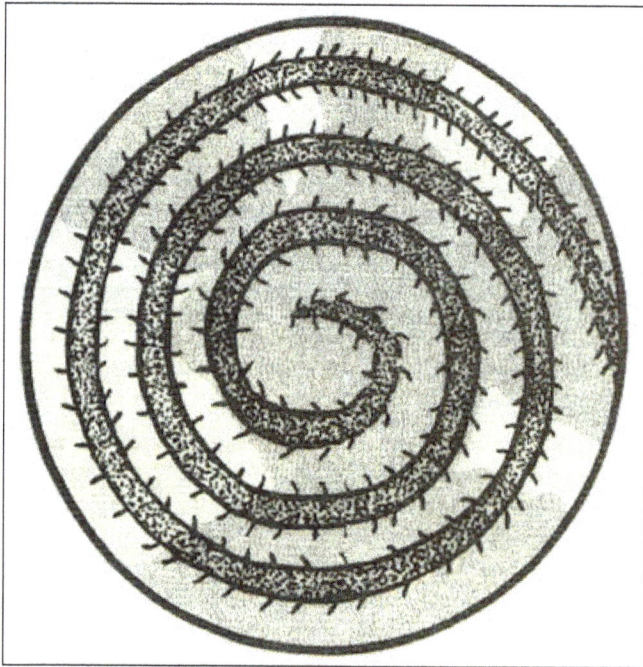

**Figure 2.6: Spiral Representing Peyote.**
**Image taken from Marco Antonio Pacheco, Raíces (Arqueología Mexicana, No. 59).**

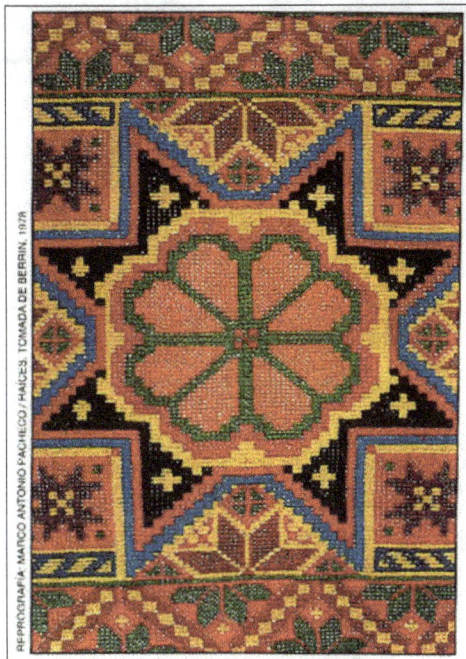

**Figure 2.7: Textile Motifs as a Representation on Shamanic Journey. @Berrin, 1978. (Arqueología Mexicna, No.59).**

**Figure 2.8: Murals that Represent Toloache (*Datura stramonium*). @Jiménez, 1998 (Arqueología Mexicana, No.59).**

## 2.6 Testimonial Iconography

According to Rodriguez Cabrera (2003), Iconography is another supportive argument for the usage of hallucinogenic beverages in ancient Mexico. There are Pre-Hispanic murals (Figure 2.9), codices, and vases that depict clearly the consumption of intoxicating beverages in social and religious contexts. Codices *Dresden, Borgia, Borbónico*, and *Vindobonensis*, among others, are the main sources of images showing how these beverages like *pulque* were consumed. Some important scenes are found in mural paintings like in Atetelco, Teotihuacan where individuals on a pyramidal platform were dancing, eating, and drinking probably a hallucinogenic beverage.

Another interesting mural to be mentioned is the one located in Cholula, Puebla, known as "The Drinkers". The figures present there are celebrating a ritual. It is suggested by Lizardi-Ramos (1969) that they are drinking *pulque*. One of the figures is carrying an *acocote*, an artifact use to extract *aguamiel* or sap from maguey plants (Müller, 1972). There is an insect, a large vat or *tlachiquera* (for collecting *pulque*), as well as glasses similar to those found in offerings and tombs. One figure is pouring a white liquid, while another wears a mask, possible representing a rabbit, an animal associated with *pulque*, agriculture, and the Moon.

Murals can be related to records of ritualistic aspects for *pulque* drinking. They were places in public view and could represent a way of communication system

**Figure 2.9: Mural Painting in Teotihuacan, Mexico, which is a
Representation of a Hallucinogenic Plant Associated with a Goddess.
@Marco Antonio Pacheco/Raíces (Arqueologia Mexicna, No. 59).**

that could be used to reinforce religious thought to the community. These murals illustrate a society, which was in continuous contact with the gods and ancestors to declare its legitimacy.

## 2.7 Rituality for Psychoactive Plants

Even psychoactive substances are sometimes regarded as drugs in Western culture, many other societies all over the world and through history, such substances have been used as religious instruments for rituals and spiritual work (Shanon, 2002). From ancient times, primitive societies have practiced the study of self-awareness. They examined the connection of their senses, mind and body to develop self-consciousness by searching for God. This is so called entheogenesis. In reference to this concept, the term entheogens can be defined as agents that bring one in touch with the Divine within (Shepard, 2005; Shanon, 2002; Ott, 1996; Ruck *et al.*, 1979) recently coined word referring to hallucinogens that induce mystical experiences. Nowadays, psychoactive agents are more generally known as psychedelics (etymologically, mind manifesting) or hallucinogenic instead of the term *entheogens*.

There are other terms used like 'plants of power' or 'plants of the gods.' In any case, none of them fully describes the broad range of effects that psychoactive plants have on the human mind and body, including visual and auditory hallucinations, other sensory distortions, mood alterations, enhanced social interactions, personal and spiritual insights, bodily purging, physical and psychological healing, and in some cases, mystical or ecstatic religious experience (Shepard, 2005).

Entheogenesis, a cultural practice still done by indigenous cultures in America, is a primary function of the shaman, who, through trance or ecstasy, is able to communicate with the dead, the gods and demons or spirits of nature and become the mediator between the human world and the supernatural realm (Aguilar, 2003). In order to do this communication, shamans ingest psychoactive substances that can have an effect in humans. Psychotonic substances cause mental excitation; psychodisleptic substances reduce tension, inducing drowsiness and cause illumination. The latter term means according to Aguilar (2003), a state of mind that generates hallucinations with bright colours, reduce perception of space and time, sensation of ecstasy, internal peace, universal and fraternal love, introspection, a sense of the recovery of past memory and a feeling of intimate union with nature and oneness with the cosmos.

Like in American cultures, in ancient Europe, mushrooms were also used for rituals such as in Greece, where psychotropic mushrooms were dedicated to Demeter, the goddess of fertility, but in Medieval times they were consider demoniac. In India, mushrooms like *Amanita muscaria* were sacred in the Rig Veda and in Egypt, mandragora (*Nymphacea caerulea*), *Datura* and henbane both were used for the same purpose.

In Mexico, psychotropic mushrooms were used and has continued to the present day. For instance, Mazatecs shamans used them to cure diseases and for divinatory purposes (Benitez, 1964). Maya shamans (*Ah men or Chilam*) also had a huge spectrum of plant material in which entheogenic substances were used to cure psychosomatic diseases. That material included hallucinogenic mushrooms such as *Amanita muscaria* (containing muscarine and ibotenic acid); *Psilocybe genus* (containing psilocin and psilocybin); white lotus flower (*Nymphaea ampla*), and tobacco (*Nicotiana rustica*). Those substances have powerful psychoactive effects. It is interesting to point out that in *Maya* codices (Dresden and Madrid), mushrooms appear in scenes with human sacrifices too for their religion. The ingestion of those mushrooms allowed shamans to have access to the realm of *Chac*, who had the capacity to control rain. By using ecstatic state and divination, they were able to communicate with the forces of nature and becoming an embodiment of Corn and Rain. The white lotus flower, an hallucinogenic plant related to chain of fertility represented on how the flowers fed the fish, fertilized the soil and finally access for the cultivation of corn; this flower was in the headdresses of *Maya* rulers and became a symbol of a lineage. In sculptures, this plant emerges from different parts of the body also symbolizing death. This could have the meaning of the psychotropic effects as seen in Codex Dresden, where Chac pulls this flower from the water. In addition, in recent *Maya* communities like *Lacandon*, another ritual is present. A beverage called balché is used for agricultural ceremonies as a type of libation. This is prepared with honey, Cenote's water, and sundried bark of the balché tree. The bark is added to aguamiel (*maguey* sap) and submitted to fermentation in earthen containers (Aguilar, 2003). For Teotihuacan, mushrooms appear together with the *ololiuhqui* plant or quiebra cajete blanco (*Turbina corymbosa*) in its murals (Lozoya, 1999; Furst, 1974).

For Aztecs rituality, hallucinogenic mushrooms were called *teoanácatl*, 'flesh or fungus of the gods.' These hallucinogenic mushrooms were represented on the sculpture of Xochipilli, the lord of psychedelic flowers. This deity was in ecstasy, where the carving on his body also included tobacco flowers, white *quiebra cajete*, *sinecuichi*, and *poyomatli* flowers. Along with this deity, Aztecs had another one known as *Pilzintli*, *Pilzintcuhtli*, *Pilzinteotl*, or *Teopiltzin*, meaning the Child-God (*Niño Dios*), who is the patron of the 'little people' or 'holy children', both names for hallucinogenic mushrooms (Aguilar, 2003).

In Guatemala, Niño Dios (Child-God) of Amatitlan, a Christian version of 'Santo Niño de Atocha,' is still worshiped by Pokoman Maya. Considering this statement, this image is still connected with Child-God of the indigenous mind, Pilzintli, due to its presence in a decorative manner in Immaculate Conception Church in Antigua, and in Puebla, Mexico in Santa María Tonanzintla. Apart from that, fungus-shaped stones found in Amatitlan Lake sustain this argument (Aguilar, 2003).

*Datura* or *Toloache* (*Datura stramonium*) is considered an important medicinal and ritual plant for many ethnic groups in northern and central Mexico as well as in some groups from southern California. It is used for acquiring visions, as an amulet to win bets, for diagnostic purposes, as an aid in hunting, and in the initiation rites of youths of both sexes. For instance, the Seri think that *datura* has supernatural powers and have an invisible spirit inside it. They also use it as a mean to do weather manipulations, cleansing of people, make fetishes and cure some type of illnesses. For shamans of *Pima oh'dam* ethnic group in Sonora and Arizona, this plant is used for visions and as a helper for hunters to find their prey. Other conceptions are determined in the mind of Huichol for *toloache*. They called it *kieri* and it is the opposite of *peyote* (*jíkuri*). In a mythological battle, *peyote* gets the victory over *kieri*. Despite *kieri* is less used than *jíkuri*, it is seen as an ally of shamans (*mara'kames*), and musicians. The corn, consider as a female plant for the Tepehuano, is an important plant where *toloache* is its husband and the son-in-law of the Sun. Toloache had two lovers and it was punished by obliging it to lower its head and fulfill the desires and whims of whoever requested its services. By doing comparison of the previous statement, love becomes an important topic, where *toloache* means a plant that is used as love amulet and when given to someone, it can fall in love, go crazy or be left in a daze (Ramírez, 2003).

## 2.8 Shamanic Journeys

In terms of Eliade (2003), the concept of shaman can be typically confused and used with other concepts such as medicine-man, magician or sorcerer. Obviously, shaman itself is too a wizard and a medicine-man at the same time because of its capacity to cure like a medical doctor and do miracles as magicians do, but it is also a psychopomp and could be a priest, mystic or poet as well. It plays many roles in one.

Another definition of a shaman is a person who is a religious practitioner and acts as an intermediary between the 'natural' and 'supernatural' worlds or sacred spheres. Cross-cultural research has indicated that shamans typically undergo a shamanic journey. The Classical Shamanic Journey has three phases. The first one is when the shaman's spirit sets forth from the natural world, then the following step it continues with the journey to the world of the supernatural where it interacts with spirit beings to gather knowledge or ensure favours, *i.e.*, rain, and the final step is when shaman returns to the natural world and re-enters the shaman's body (VanPool, 2003).

The shaman conserves the capacity to establish direct relationship with the spiritual world that it is reflected in the peculiar interaction with the natural environment. When the shaman is entirely in a state of trance, travels to the inner earth or sky to gather and collaborate with deities or spirit helpers. Once the shaman enters to the alter state of consciousness (arrives the other world), starts its communication with animals spirits or gains fights against malevolent influences. After this process, the shaman returns to the everyday living world showing its ability to predict the future to help others to solve their problems like emotional difficulties or other illnesses.

Generally speaking, a shaman starts as oneself when one suffers an abrupt and mysterious illness, an attack of mental imbalance, and sometimes, after having unusual and intense dreams. A new beginning becomes evident after realizing that the shaman itself is able to submerge in an ecstatic state of trance and develops special abilities throughout different practices. Some shamans enter to the spiritual world by the abstinence of dreaming, water ingestion or food, as the algonquins do. Others, like matsigengas, look for the sky and world of the deaths by ingesting a beverage containing a hallucinogenic plant called Ayahuasca in Peru. Some others access to these types of powers from inside of a cave where voices of animals like bears, pumas or others call from the inner zone of a mountain. In some regions of Africa, a shaman is subjected to several months for training staying isolated in cult places. After the shaman finishes, he uses another name and get out transformed that reflect his new condition and social category (Leigh-Molyneaux, 2002).

A distinguished example of a shamanic journey is shown by the pottery of Casas Grandes, Chihuahua, Mexico, dated from 1250-1450, A.D., and includes images of smokers, dancers, and humans with macaw heads that have two designs, pound signs and circles with dots. As VanPool (2003) described it, the first phase of this journey starts when the shaman smokes the pipe probably with tobacco to induce a trace, enabling him to travel to the spirit world. An example of it can be seen in murals from Maya region (Figure 2.10). Later one, he dances to further continuing the trance induction starting the transformation. Finally, he changes into a spirit being.

Parallel to the previous concepts, shamanism is a Siberian and Central-Asian phenomenon. The word comes from *tungus, shaman*. In *tungus* means "the one who can see" or "someone who knows the ecstasy" as mentioned by Mercier (2009). In other languages from North and Central Asia the correspondent terms are: in *yakuto, ojun*; in *mongol, bögä*, and *udagan, udayan* – shaman female; and in turkish-tataro, *kam* (Eliade, 2003).

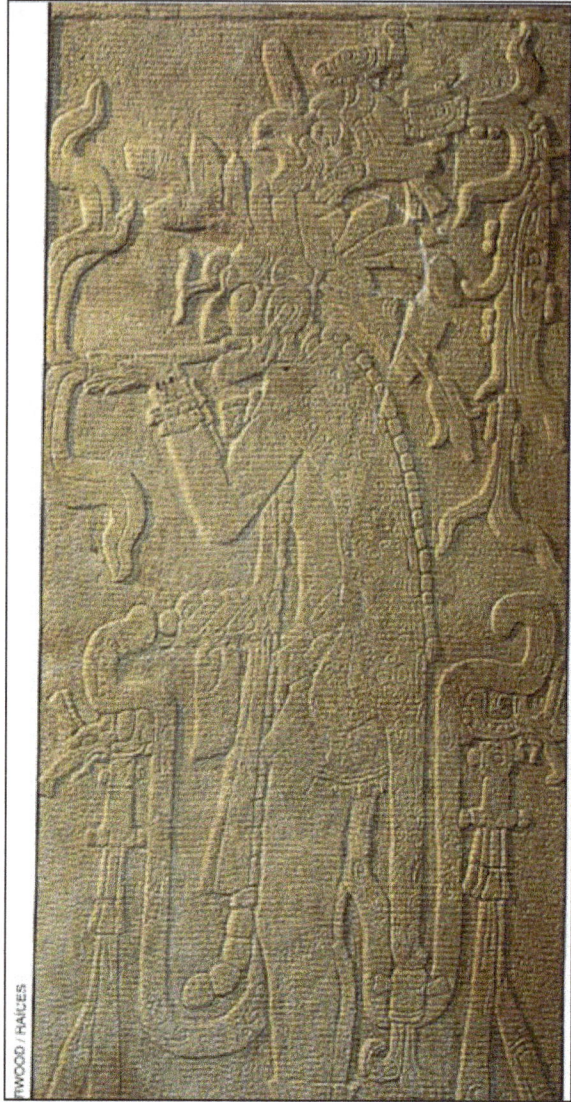

**Figure 2.10: Mural (Relief) Representing Snuffin for Shamanic Journeys.**
**@ Michael Calderwood. Raíces.**

In all Central and Northern Asia, magical-religious life of this society revolves around the shaman, so not every single aspect is manipulated by it. There are other important figures that play a role in the sacred activities. In tribes, the priest coexist with the shaman and the head of the family is the head of the domestic cult. However, shamans are considered the dominant figure due to its ecstatic experiences is a religious experience where the shaman is the most knowledgeable and is considered by the group as the master of the ecstasy. Shamans are specialists in trances where it is believed that their soul abandoned their body to undertake a rise to heaven or fall to hell. Another particularity is that either modern or primitive shamans believe that they have connections with helper spirits by being possessed by them or simply dominate them entirely for their convenience.

In respect to how shamans are selected for this profession or work, there are three ways. One is that they receive knowledge transfer by inheritance; the second is by spontaneous vocation or gods or spiritual calling, and the third one is for own will. The last one is considered the weakest one from the other, and the strongest would be the spiritual call from the gods. An important detail for the election of being a shaman is that it matches almost always with a manifestation of a very serious illness. Even though a shaman can be chosen for the previous three forms that were explained before, it can be recognized as such when a double instruction is done, which is equivalent to the term initiation. This means that the instruction is done by dreams or trances and traditionally by shamanic technics, name of the spirits, mythology, secret language, and clan genealogy (Eliade, 2003).

Apart from the historical relationships between Asia and America, in terms of shamanism exist the same features and cultural elements among the ethnic groups. The process of initiation well describe by Eliade (2003) indicates that this phenomenon is present everywhere in the American continent with the same great traits of initiation in the mystical life: vocation, retreat to solitude, learning and understanding the teachings of a master, acquiring one or more familiar spirits, do symbolic ritual of death and resurrection, use of secret language. All of these are elements that identify in a sense why a shaman journey must exist.

## 2.9 Hallucinogens in Contemporary Cultures

Since old times, human societies have used stimulants as therapeutics in their cultural processes. By using them for a long time and systematically for being dynamic inside their culture, it is interesting to note that this long process was acquired through trial and error.

More recently, concepts such as hallucinogens or alkaloids are discredited everywhere for their illegal use in Western countries (Masferrer Kan and Díaz Brenis, 2000). Nonetheless, tobacco and caffeine, which are in a big demand and are used intensively by human societies everywhere in the world, contain alkaloids and are not considered as such.

Wisely indigenous groups have social control in the excess of stimulants consumption by establishing a set of strategies using complex ritual systems, in which they include these substances into their daily life by encouraging the dynamism of everyday activities. Their social development is completely different and cannot be compared with the addictive use of modern societies, which destroy and worsen psychopathic behaviour. An interesting example of the previous statement is the use of *coca* leaves in the Andean zone. What they do is to take away the rib of a leaf before chewing it, using a pinch of lime to enable the release of the alkaloid by using their saliva. By chewing *coca* leaves, indigenous people consume less than 23 g of the leaf per day in comparison to the production of one gram of cocaine chlorhydrate, which requires 8 kg of leaves to be processed for addicted people. *Coca* leaves are sold freely in Peru, Bolivia, and northern Argentina. Same case for *Ayahuasca* (vine of souls, *Banisteriopsis caapi*) which has a legal status in Brazil in Santo Daime's Church, where is used for worshipping. In Peru, the Ministry of Health has authorized the Centro Takiwasi de Tarapoto as a rehabilitation centre in with an interdisciplinary team consisted of doctors, psychologists, and Amazonian shamans that assist people using *ayahuasca* for therapeutic purposes (Masferrer Kan, 2003).

Mexican special case is the use of *peyote* and several hallucinogenic mushrooms, which are broadly used as stimulants. Peyote is a small spineless cactus that grows in the desert of most of the northern and central part of Mexico and hallucinogenic mushrooms are native to moist, subtropical areas like Oaxaca, Chiapas, and Puebla. In Huautla de Jiménez, Oaxaca, Mazatecs like María Sabina, became famous by the use of hallucinogens and their miraculous cures done as part of their cultural activities. Many foreigners assisted to experience those activities with them, as the Mazatecs treat them for detoxification and contamination of modern society. The flux of foreign people altered the traditional lifestyle of indigenous people (Masferrer Kan, 2003).

By now, it is known that Vienna Convention Against Illicit Traffic in Narcotic Drugs and Psychotropic Substances allows native groups that use alkaloids to be part of their traditional cultural baggage. Nevertheless, hallucinogens and stimulants have been and will continue to be present in societies worldwide, each individual society must define how they need to be consumed in order to allow them to serve as a dynamic, positive element and enriching therapeutic resource instead of being a big problem for the society.

## 2.10 Ethnobotanical Approach and Exploration

Plants are fundamental for every existing life organisms on earth. They have the capacity to absorb carbon dioxide and expel oxygen through photosynthesis, which is the basis for food production for all organisms. In addition, they reuse nutrients, stabilize land erosion, and control atmospheric water cycle via plant transpiration (Cotton, 1996).

Medicinal plant use has a long history through time which has more that 4000 years old according to documents from old civilizations in China and India (Hamburger and Hostettmann, 1991). Approximately 64 percent of the world population depends on traditional medicine (Cotton, 1996) and 25 percent of regular medicinal products are derived from plants (Hamburger and Hostettmann, 1991).

In plant kingdom exists approximately 500 000 species from which a small amount of them has been investigated for their phytochemical compounds. Medicinal knowledge and their pharmaceutical and biological properties are very scarce. These plants contain high amount of secondary metabolites, so there is an enormous potential to find more biological activity from other compounds (Hostettmann *et al.*, 1995).

Ethnobotany, as a whole, is a link between plants and people. Its definition implies the botanical knowledge of plants from indigenous communities and a close and strong relation between them (Cotton, 1996).

Plants have been a rich source of medicines because they produce a host of bioactive molecules, most of which probably evolved as chemical defences against predation or infection. Many chemicals that are toxic to insects or predators also show bioactivity in humans, which means they might be capable of achieving some therapeutic effects. From the previous sentence, it is a must to understand the meaning of what Sharma (1996) states: 'the proper use makes a good remedy even out of poison if used improperly'.

Thus, ethnobotanical approach assumes that indigenous uses of plants can offer strong clues to the biological activities of plants; and the history of drug discovery definitely implies that the ethnobotanical approach would be the most productive of the plant-surveying methods, and recent findings confirm that impression. So far the rigorous ethnobotanical field searches, which have been conducted since the early 1980's, have generated many lead compounds. Some of these have been identified as drug candidates and most of them exhibit potent antiviral, antifungal or anticancer effects (Cox and Balick, 1994).

A new integral approach of research was generated by unifying different specialties with each other and coordinated them with data generated from each discipline in relation to the other to better understand plants used traditionally. This was manifested particularly in Mexico with the incursion of Schultes throughout his research on sacred plants and the ethnobotanical research implicated on it during 1970's. From this point of view, other disciplines such as ethnopharmacology, which is defined as a scientific discipline that searches bioactive substances from plants (Farnsworth, 1994), and other related areas were adding a more comprehensive material to study and understand the active principles in correlation with the traditional uses and their effects on humans. It was necessary to define every aspect of each discipline to clearly understand the plants themselves. As a result, traditional uses such as rites and beliefs of sacred plants were the link to interconnect the perception of the mental effects caused by the sacred plants and their molecular compounds associated with them, forming the basis for this integrated research. By getting such an interesting methodological approach with connections between every discipline, psychoactive plant taxonomy for traditional used plants of Mesoamerica was created. This system takes into account a more detailed and comparative analysis of such diverse elements as traditional uses, mental effects and their chemical compound's structure. Louis Lewin, a German toxicologist, who was one of the founders and users of the psychopharmacology as an interrelated discipline, made a classification of five groups of psychoactive drugs in a book from 1924, so called *Phantastica*. In it, he emphasized on the main effect of visual characteristic of psychotropic plants, the increased visual imagination, as the five titles were denoted as: *Exitantia*, stimulant plants where coffee, tea, and yerba mate were classified in this group; *Euphorica*, for euphoriant plants, opium and coca were selected; *Inebriantia*, inebriating drugs that are fermented or distilled were mentioned; *Hypnotica*, a somniferous drugs like chloral; and the last one,

*Phantastica*, for hallucinogenic plants like peyote as one of the main sacred plants included (Díaz, 2003).

In regards to the visual characteristics of perception of the sacred plants when ingested, an important one is the manifestation of geometric figures mixed with intense bright colours mimicking their action on visual effects; and the other one, which is the main and well known feature, is the magnification and intensity of the imagination and fantasy due to their neurotropic molecules. Cultural manifestations are presented and expressed in murals on stones and petroglyphs, which were made by indigenous peoples such as Huicholes, as a result of these visual perceptions.

Nowadays, interdisciplinary research teams all over the world have been studying psychoactive and psychotropic substances extracted from plants, as do biochemical studies. This intensive work has to do with the development of new ways of developing good therapeutic treatments for drug addiction and mental illnesses like depression. Interesting research work of plants such as *Tabernanthe iboga* (Mash *et al.*, 2000) and *Banisteriopsis caapi* (Palladino, 2009; Thomas *et al.*, 2013) are considered to be very important in this field (Sobiecki, 2014).

In addition, there are some teams that had developed research and educational organizations such as Multidisciplinary Association for Psychedelic Research (MAPS) which its main priority consists of a variety of research topics related to psychoactive plants being the main ones: MDMA-Assisted Psychotherapy, LSD-Assisted Psychotherapy, Ibogaine-Assited Treatment, Ayahuasca-Assisted Therapy, and Medical Marijuana.

In the indigenous world, magical and religious uses of psychoactive plants play an important role in the community for rituals and, at the same time, for medicinal purposes.

From the above, it can be said that ethnobotany should be considered an art by itself because it is based on different areas of knowledge in concordance with problems, as stated by Hernández (1970). It is an intellectual and material bridge between the indigenous person and the rest of the professions involved on the problem such as biochemist, physiologist, medical doctors, agronomist, genetists, historians, linguists, and many more.

## 2.11 Ethnobotany and Ethnomedicine in southern Chile

Shamanism in Chile is very broad. In southern regions, it is expressed for several cultural groups. In our case, *Williche* communities that are part of the genera of *mapuche* family have their own construction and believe that were accumulated and acculturated by *mapuche* influence. *Williche* culture carries a rich broad of beliefs and a good variety of rites, which allow men to put themselves in contact with the forces of nature and the supernatural world. The *machi* (*shaman*), who keeps contact and mediates with these two worlds, plays an important role in the cosmological system.

The moon (*Killén*), the morning star (*Wuñelf*), and the stars (*Wanglén*) are deities who influence the *machi* whose magical gifts depend on these celestial bodies. Prayers are done by *machis* to solicit the intercession of ancestors and fallen heroes who have reached mythic heights, invoking the spirits of past warriors, chieftains and revered *machis* to look for the safety and prosperity of their descendants, as they did in life. Natural forces, which are linked to various believes have a mythic connotation to the land's parts.

During some ceremonial prayers such as *ngillatun*, the anthropomorphic figures (*nguillatue*) are also oriented towards the eastern mountain; their line of sight must remain unobstructed during the ceremony. The *machi* installs her *rewe* (ladder to heaven) in a manner in which looking at the statues, she directs her prayers to the East. In this ritual, plants are of the most importance for the ceremonial prayer done by *machis*. The *folie*, or cinnamon tree, is worshipped as the embodiment of divine attributes and messenger of peace. The *maqui* (*Aristotelia chilensis*), and the *laurel* (*Laurelia sempervirens*) also assume these characteristics, and their use is common in decorative religious places and objects for *machi* prayers. The *machi* or *fileu*, is the intermediary between the *mapuche* people and the *wenu mapu* (land of the gods). Divinities grant health, well-being, abundance for the *machi* who is charge with divine and has the power to cure and conduct rituals. She usually has a struggle between good and evil on earth (Museo Chileno de Arte Precolombino and Ministerio de Educación, 1986).

There are many hallucinogenic plants that were discovered over the time by Williche people. They are important for magical-religious treatments. Indigenous people from this area used to drink fermented beverages from endemic fruits. These beverages play an important role in religious practices (Mösbach, 1999; Hoffmann, 1997). They used to prepare them using the following fruits: *Luma, maqui, michay, frutilla, lleuque, murta*. Even though, some of these beverages still remain as a cultural practice in terms of preparation, such as *cauchao*, a fruit from *luma* (*Amomytus luma*) or *maqui* (*Aristotelia chilensis*), some others are just consumed as a fruit. There are other plants that are particularly used for ceremonials conducted by machis and some of them are listed below and are mentioned with their relevant information about religious and magical characteristics that are well-known by Williche communities. All the plants listed here have other medicinal properties for modern ailments and are not added in this manuscript. Not all of them are used by machis but have magical properties that can cause an effect on the psyche of a person.

## 2.12 Plants with Magical and Religious Uses by Williche People

**Chaura**, *Gaultheria phillyreifolia* (Pers.) Sleumer, Ericaceae. Bush (Figure 2.11). The whole plant or branches with thorns are used to beat softly the head and other part of the body to get rid of bad spirits that provoke spiritual ailments. It is also used to 'clean' places such as houses or agricultural lands to get prosperity and good harvests by doing smoke-offering. The sounds produced when firing can expel the supernatural entities.

**Latúe**, *Latua pubiflora* (Griseb.) Baillon, Solanaceae (Figure 2.12). An endemic bush use by sorcerers. It is a very toxic plant that can cause death, if ingested. Its usage is restricted to machis as a hallucinogen. It is a powerful plant, consider by the community as alive. Considering its magical properties, it can benefit or cure a sick person, give prosperity to a family, have a good planting and harvest (Jeréz, 2005; Montecino, 2003, Olivares, 1995). Another use generally accepted by Williche people is that brujos (sorcerers) employ this plant to send bad energies to someone.

Figure 2.11: Chaura, *Gaultheria phillyreifolia* (Pers.) Sleumer, Ericaceae. Bush.
@ Roberto Saavedra.

Figure 2.12: Latúe, *Latua pubiflora* (Griseb.) Baillon, Solanaceae.
@ Roberto Saavedra.

In addition, some others argue that latue is a friendly plant acting as a helper. It is possible to talk to it and ask for some favours. It is also believed that this plant only grows close to someone who can cure sick people with plants. Its leaves and fruit are very toxic, so if a person ingest any part of it, or particularly the fruit, could immediately get very crazy. Thus, it is necessary to ingest another plant (yagui) agains it to reestablish.

**Salvia**, *Sphacele chamaedryoides* (Balbis) Briq., Lamiaceae (Figure 2.13). Herbaceous plant. It was documented that this plant is effective against bad supernatural entities by getting smoke-offerings to clean places and people.

**Copihue**, *Lapageria rosea* Ruiz et Pavon. Philesiaceae (Figure 2.14). Climbing plant use as a harbinger for machi election and future vocation (Citarella *et al.*, 1995). It is inserted in the mapuche-williche mythology of creation and a national flower of Chile. It is salso a spirit for canelo that takes away light from sorcerer (Montecino, 2003).

**Voqui Quilmay**, *Elytropus chilensis* (A.DC.) Muell.Arg. Apocynaceae (Figure 2.15). Climbing plant. It is use for smoke-offerings to clean home atmosphere avoinding bad spirits.

**Melí**, *Amomyrtus meli* (Phil.) D.Legrand et Kausel, Myrtaceae (Figure 2.16). Endemic tree. Using some stems from this plant, a cross can be made to place it at the entrance of a house believing to have it for protection and to avoid sorceress to get into the place.

**Huautro**, *Baccharis concava* (R. Et P.) Pers. Asteraceae (Figure 2.17). It is used by Williche people for magical-religious ailments. It can be applied as a topic over some parts of the body to extract and eliminate corrientes (bad vibrations or internal currents).

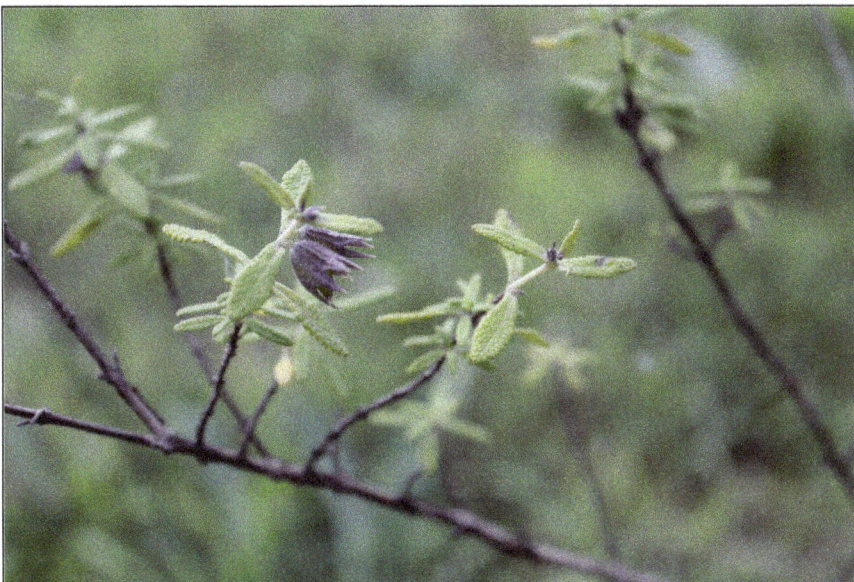

**Figure 2.13: Salvia, *Sphacele chamaedryoides* (Balbis) Briq., Lamiaceae.**
**@ Roberto Saavedra.**

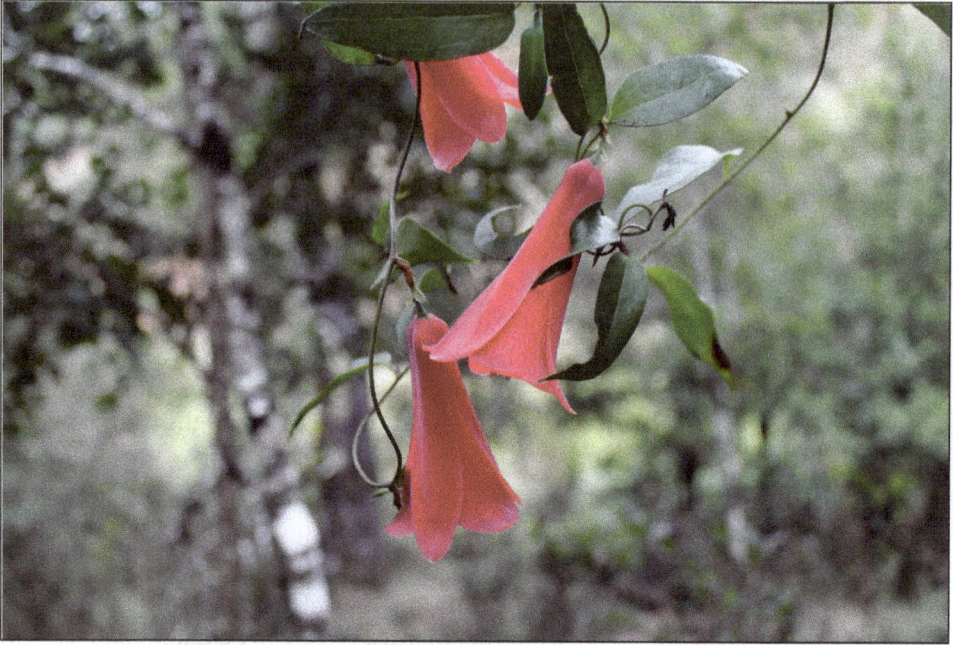

**Figure 2.14: Copihue, *Lapageria rosea* Ruiz et Pavon. Philesiaceae. @ Roberto Saavedra.**

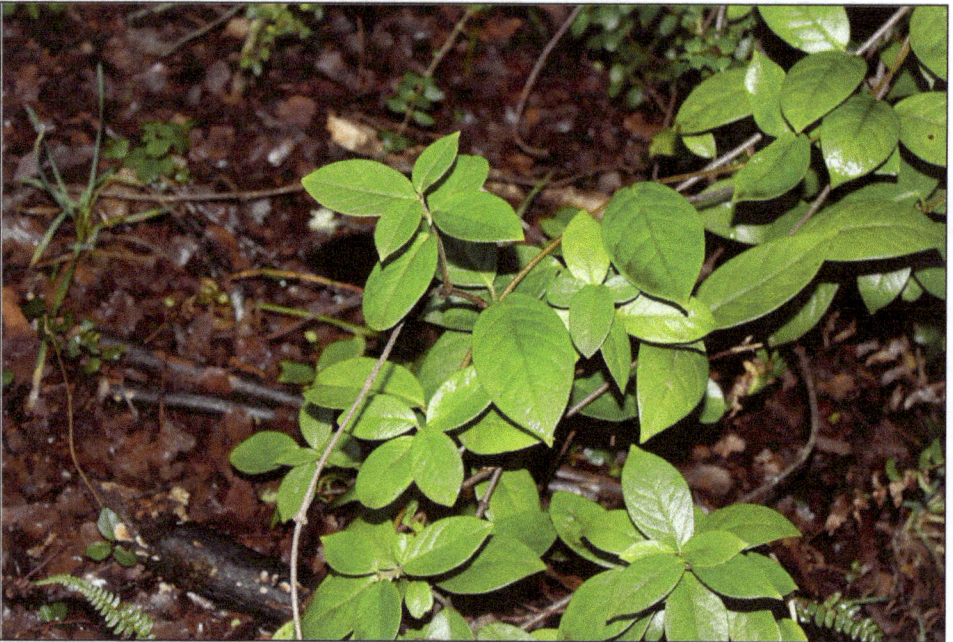

**Figure 2.15: Voqui Quilmay, *Elytropus chilensis* (A.DC.) Muell.Arg. Apocynaceae. @ Roberto Saavedra.**

Figure 2.16: Melí, *Amomyrtus meli* (Phil.) D.Legrand et Kausel, Myrtaceae.
@ Roberto Saavedra.

Figure 2.17: Huautro, *Baccharis concava* (R. Et P.) Pers. Asteraceae.
@ Roberto Saavedra.

**Boldo,** *Peumus boldo* Mol., Monimiaceae (Figure 2.18). Tree used for several ailments such as liver and stomach problems. It has an important alakloid boldin. It was found as an hallucinogen dated back from Monte Verde, an old human settlement in Southern Chile (Dillehay, 2004). *Machis* (*shamans*) use to smoke its leaves with pipes to induce trances for rituals and ceremonies (Jerez, 2005; Guevara, 2000; Rosales, 1989).

Figure 2.18: Boldo, *Peumus boldo* Mol., Monimiaceae.
@ Roberto Saavedra.

**Coicopihue,** *Philesia magellanica* J.F. Gmel, Philesaceae (Figure 2.19). Climbing bush. By tradition, it is believed that it florishes when Trauko, a mythical entity passes, closet to it. Also from the cuñeima a new born creature cries when new moon appears announcing treasuries (Jerez, 2005; Montecino, 2003).

**Trauman,** *Pseudopanax laetevirens* (Gay) Seem ex Reiche. Araliaceae (Figure 2.20). Climbing bush. It is used to get rid of bad spirits. Cárdenas (1997) indicates that it was used as a special beverage for treputo or cheputo ceremony. This ritual is historically associated with the usage of trauman along with tepa (*Laureliopsis philipiana*) previously smoked to hit the fish pens for good harvesting.

**Maqui,** *Aristotelia chilensis* (Mol.) Stuntz. Elaeocarpaceae (Figure 2.21). *Machis* (shamans) use to smoke its leaves with pipes to induce trances for rituals and ceremonies (Jerez, 2005; Guevara, 2000; Rosales, 1989). This plant is also a very

**Figure 2.19: Coicopihue, *Philesia magellanica* J.F. Gmel, Philesaceae.
@ Roberto Saavedra.**

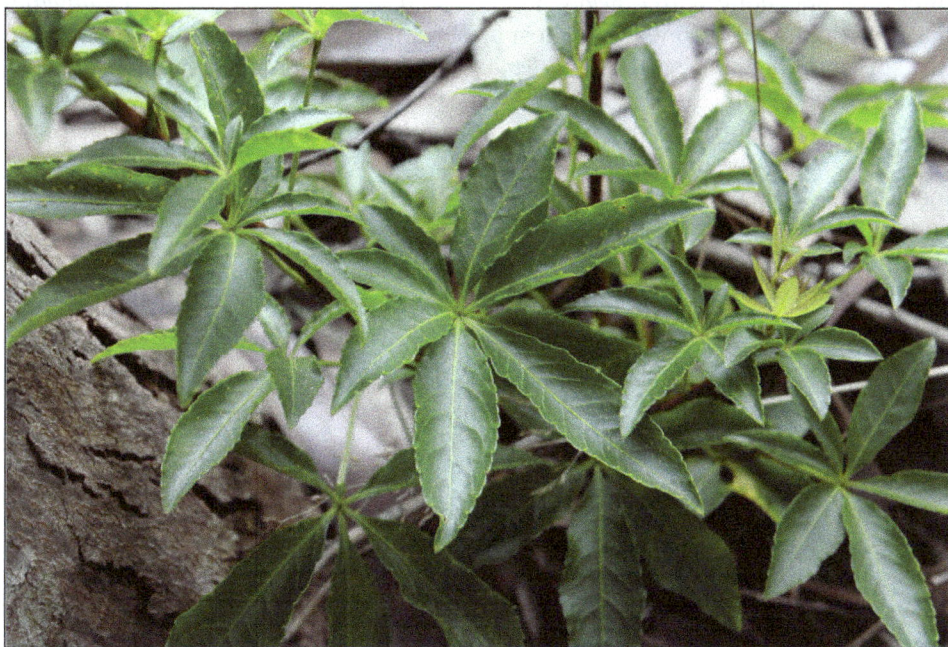

**Figure 2.20: Trauman, *Pseudopanax laetevirens* (Gay) Seem ex Reiche. Araliaceae.
@ Roberto Saavedra.**

**Figure 2.21: Maqui, *Aristotelia chilensis* (Mol.) Stuntz. elaeocarpaceae.
@ Roberto Saavedra.**

important one along with *canelo* when a special ritual *rogativa* so called *nguillatun* has been taking place (one of the most important rituals meaning in mapuzungun 'the speech of the land'; also called *lepún* (area of the ceremony). It is placed close to *canelo* aside from the *rewe* (a stair to heaven for indigenous conception). The purpose of this ritual is to have a connection with the spiritual world and ask for the strengthening of the union of the community and well-being and acknowledging the benefits received during the year such as climate, good harvesting, abundance and vitality, just to name a few examples.

**Floripondio**, *Brugmansia sanguínea* (R. et P.) Don. Solanaceae (Figure 2.22). Bush present in many places of South America. It is used as a strong hallucinogen. As a shamanic plant explores *peuma*, a dream to talk to others.

**Taique**, *Desfontainia spinosa* Ruiz et Pav., Desfontaineaceae. Bush (Figure 2.23). It was used as an hallucinogen (Schultes and Hofmann, 2012; Jeréz, 2005). This bush is consider for magical purposes and normally is avoided by the community. They consider it dangerous.

**Laurel**, *Laurelia sempervirens* (Ruiz and Pav.) Tul., Monimiaceae (Figure 2.24). Sacred tree for Williche community. It represents the blood of the ancestors and used for important rituals. It also represents peace (Jeréz, 2005; Montecino, 2003). This tree is also used for religious ceremonies to bless a festivity (a product of syncretism with *catholicism*). It can be placed at the entrance of a house or close to the main door. When a storm or rainy day is present, it is necessary to burn some branches.

Figure 2.22: Floriopondio, *Brugmansia sanguínea* (R. et P.) Don. Solanaceae.
@ Roberto Saavedra.

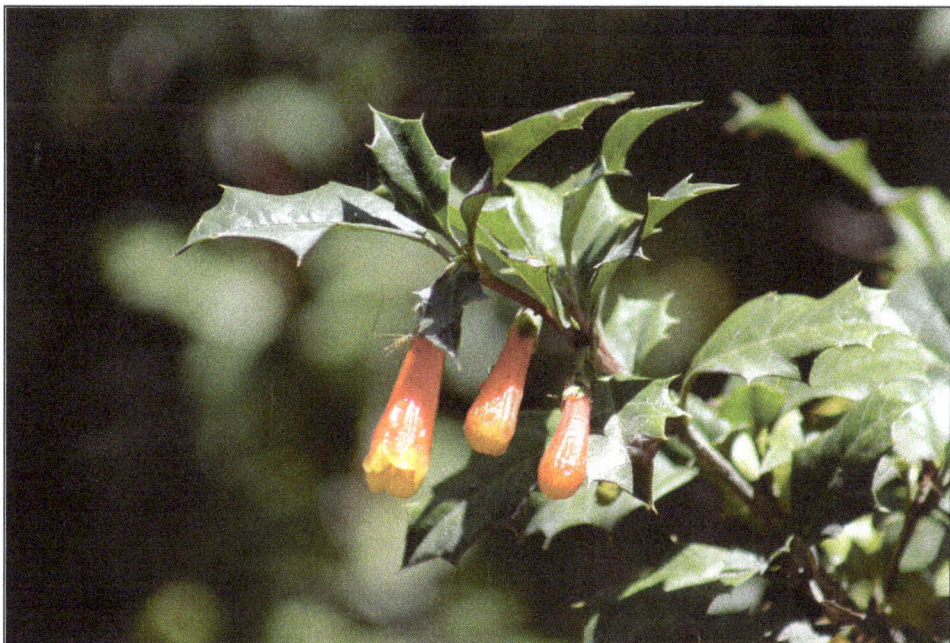

Figure 2.23: Taique, *Desfontainia spinosa* Ruiz et Pav., Desfontaineaceae. Bush.
@ Roberto Saavedra.

**Figure 2.24: Laurel, *Laurelia sempervirens* (Ruiz and Pav.) Tul., Monimiaceae.
@ Roberto Saavedra.**

**Palqui,** *Cestrum parqui* L'Hérit, Solanaceae (Figure 2.25). It is employed to attract love or to break a curse (jeréz, 2005; Cárdenas, 1997). This plant has an importance against bad energy or spiritual presences at home or other place. It is recommended to place some branches of it close to the pillow when someone go to bed and want to sleep peacefully. Besides, this plant is believed to act as a protector. For that, it is necessary to get from the plant a stem cutted in parts and elaborate a cross with them. It is believed in Williche communities its effectiveness.

**Canelo,** *Drimys winteri* J.R. et G. Forster, Winteraceae (Figure 2.26). One of the most or possibly the most important and iconic sacred tree for mapuche and Williche people. It is used during war or peace and employed for nguillatun ceremony (religious ceremony to thank for the benefits of the land and for the well being and union of the community). Appears as a shamanic symbol for machi who has the power to related with the sacred (Jeréz, 2005; Mösbach, 1999). It is used for smoke-offerings to clean home atmosphere, avoiding bad spirits. If it is ingested as an infusion, it can expel bad energies, all of them, in fact. As being one of the most important sacred trees for this indigenous group, it is necessary to say that during the *nguillatun*, a ritual explained before, several plants such as laurel, maqui, pear or apple trees are used along with it to worship with a rewe (the altar) looking to the East side in *nguillatunes*, which are carried out in a *ngillatuwe* (place specially arranged for this purpose).

**Chilco,** *Fuchsia magellanica* Lam., Onagraceae (Figure 2.27). Bush used to avoid the spiritual entity, Trauko (Jeréz, 2005; Cárdenas, 1997).

**Frutilla,** *Fragaria chiloensis* (L.) Duchesne, Rosaceae (Figure 2.28). The leaves of this plant are placed at the door to avoid entrance of a bad entities. It is used as a protective plant for its nice smell when fruits are present.

Figure 2.25: Palqui, *Cestrum parqui* L'Hérit, Solanaceae.
@Julio Larenas.

Figure 2.26: Canelo, *Drimys winteri* J.R. et G. Forster, Winteraceae.
@ Roberto Saavedra.

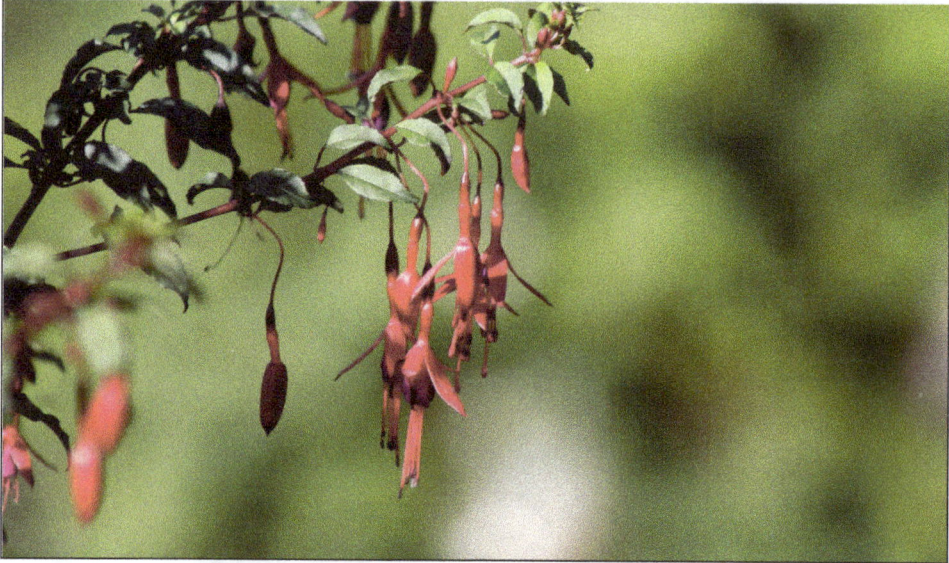

Figure 2.27: Chilco, *Fuchsia magellanica* Lam., Onagraceae.
@Julio Larenas.

Figure 2.28: Frutilla, *Fragaria chiloensis* (L.) Duchesne, Rosaceae.
@ Roberto Saavedra.

**Michay**, *Berberis darwinii* Hook., Berberidaceae (Figure 2.29). Native bush used in agricultural ceremonies in Chiloé island. A cross made of it stops culebrón, a malign entity (Jeréz, 2005; Montecino, 2003).

**Tupa**, *Lobelia tupa*. L, Lobeliaceae (Figure 2.30). Plant leaves used by indigenous people to smoke pipes to produce hallucinations (Schultes and Hofmann, 2012; Mariani, 1965). Also, the leaves are boiled with other plants and applied topically over the head to get rid of bad energies.

**Arrayán**, *Luma apiculata* (DC.), Burret, Myrtaceae (Figure 2.31). Tree that is used to cure 'aire', a non-recognized ailment by modern medicine. This means that someone is exposed to changes of temperature and different environments.

**Pil-Pil** *Voqui. Boquila trifoliolata* (D.C.) Dcne. Lardizabalaceae (Figure 2.32). Native climbing plant. It was used as a love filter and to initiate raining, calling to the spirits of the sky (ngenwenu) (Jeréz, 2005; Montecino, 2003). It is also used for rituals to get a couple united when dancing (Rosales, 1989). It has another use as a topic remedy to eliminate headaches caused by bad energies from supernatural forces.

**Matarratones**, *Coraria ruscifolia* L., Coriaraceae (Figure 2.33). Native bush which grows close to water. Machis use it for cachín, a skin ailment of magical origin (Cárdenas, 1997). It is also used against witchcraft (Jeréz, 2005).

**Figure 2.29: Michay, *Berberis darwinii* Hook., Berberidaceae.
@ Roberto Saavedra.**

**Figure 2.30: Tupa, L*obelia tupa*. L, Lobeliaceae.
@ Roberto Saavedra.**

**Figure 2.31: Arrayán, *Luma apiculata* (DC.), Burret, Myrtaceae.
@ Roberto Saavedra.**

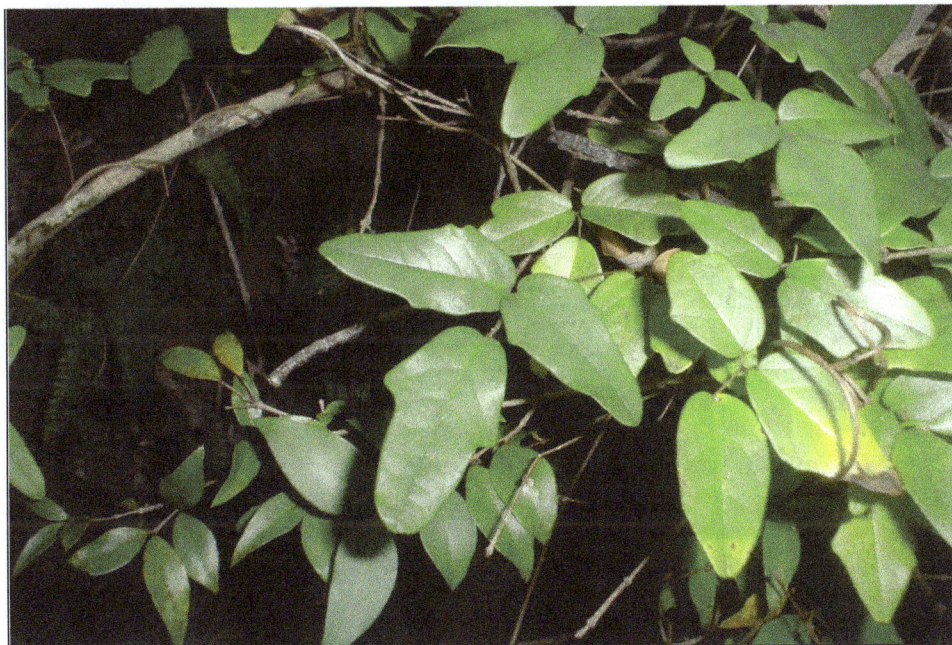

**Figure 2.32: Pil-Pil Voqui.** *Boquila trifoliolata* **(D.C.) Dcne. Lardizabalaceae. @ Roberto Saavedra.**

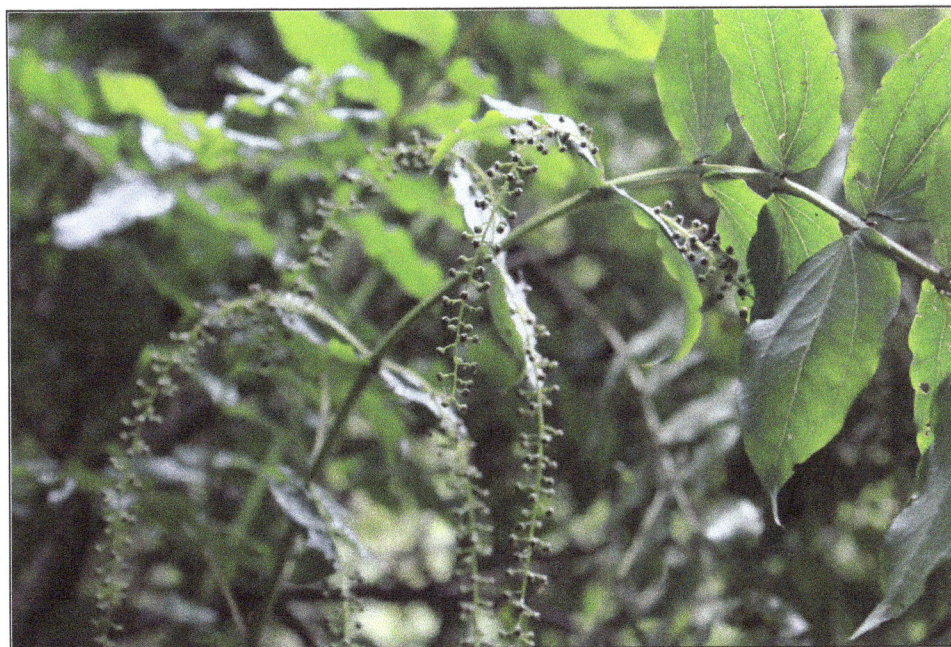

**Figure 2.33: Matarratones,** *Coraria ruscifolia* **L., Coriaraceae. @ Roberto Saavedra.**

**Quilineja**. *Luzuriaga radicans* R. Et P. Philesaceae (Figure 2.34). A climbing bush used to make crosses for tombs to avoid witches who steal corpses. Trauko, a mythical entity uses quilineja to cover his body (Montecino, 2003). This plant is also use along with other plants for cultural illnesses such as susto, a psychosomatic ailment manifested by a psychic trauma and influenced by the central nervous system. The ritual procedure for susto is unique depending on the place and the person who is in charge of it and is usually surrounded with a special paraphernalia to conduct the ritual.

**Papa**, *Solanum tuberosum* L., Solanaceae (Figure 2.35). Plant originated from Titicaca Lake. Machis (shamans) use to smoke its leaves with pipes to induce trances for rituals and ceremonies (Jerez, 2005; Guevara, 2000; Rosales, 1989). It is associated with Pleyades constellation and mapuche-williche ancestors. The parts use are the leaves.

**Llagui**, *Solanum nigrum* L. Solanaceae (Figure 2.36). Herbaceous plant that is needed when latúe is employed as a narcotic. It is considered an important plant for magical purposes in Williche pharmacopoeia.

**Quilatape**, *Griselinia ruscifolia* (Clos) Tabú (Figure 2.37). Native bush use as a love filter (Jeréz, 2005; Mösbach, 1999). Still use for extraction of bad energies in the body.

**Olivillo, Tique**. *Aextoxicom punctatum* R. Et P. Aextoxicaceae (Figure 2.38). Native tree where Trauko, a malign entity that dominates the forest, normally stays close to it (Jeréz, 2005).

**Palo Santo**. *Dasyphyllum diacanthoides* (Less.) Cabrera. Asteraceae (Figure 2.39). Native tree which is used for magical-religious purposes for Williche rituals.

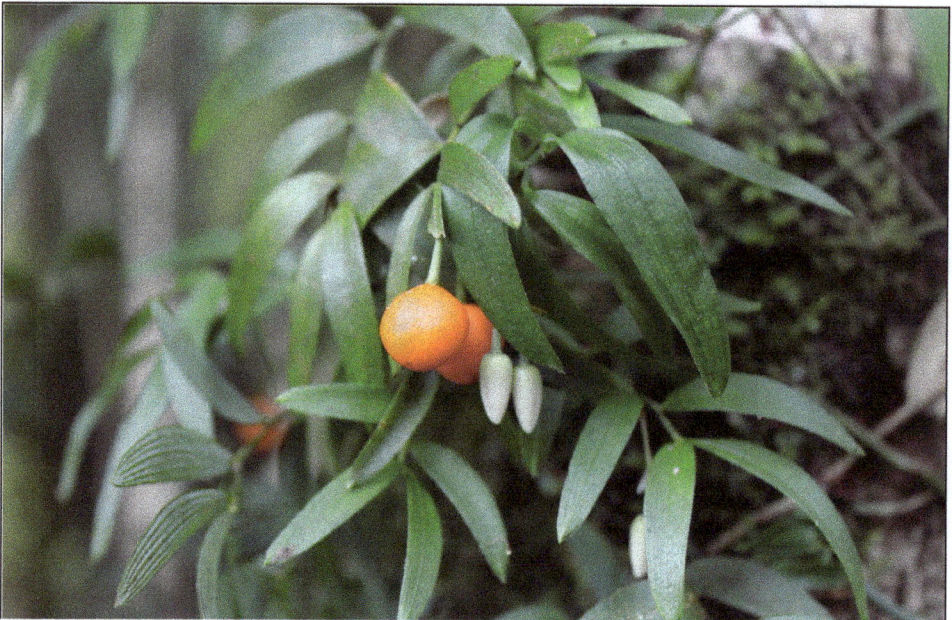

**Figure 2.34: Quilineja.** *Luzuriaga radicans* **R. Et P. Philesaceae.**
**@ Roberto Saavedra.**

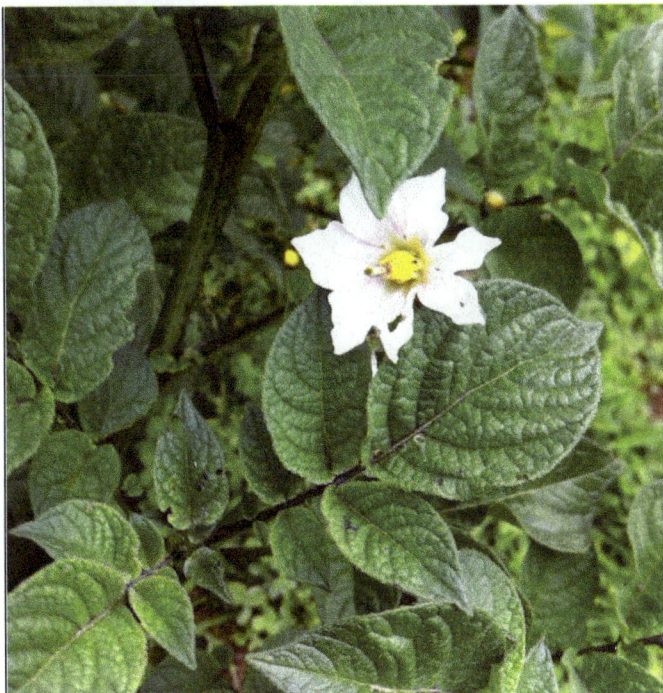

Figure 2.35: Papa, *Solanum tuberosum* L., Solanaceae.
@ Roberto Saavedra.

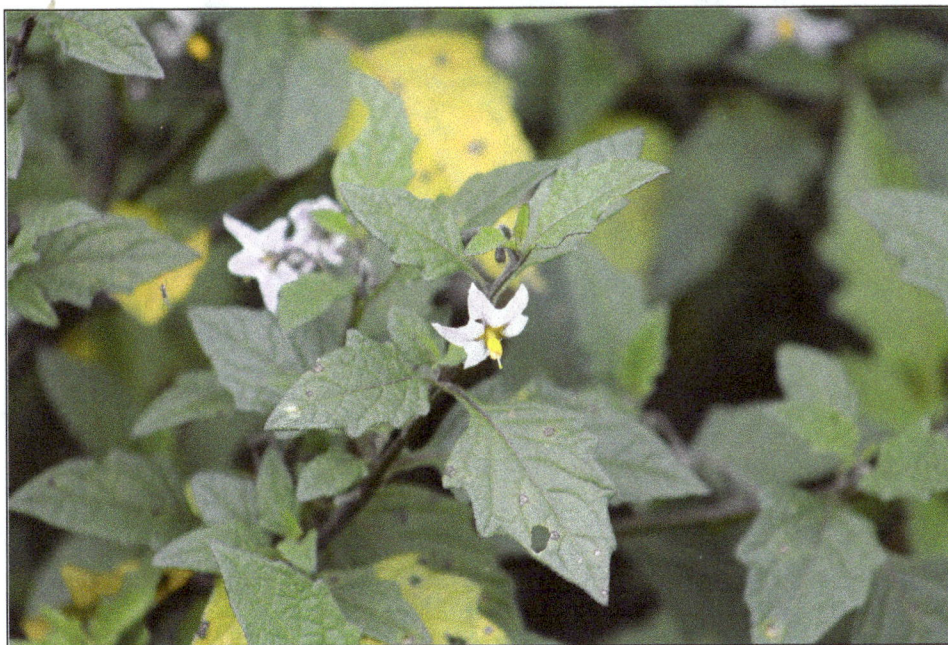

Figure 2.36: Llagui, *Solanum nigrum* L. Solanaceae.
@ Roberto Saavedra.

**Figure 2.37: Quilatape, *Griselinia ruscifolia* (Clos) Tabú.
@ Flora SBS.**

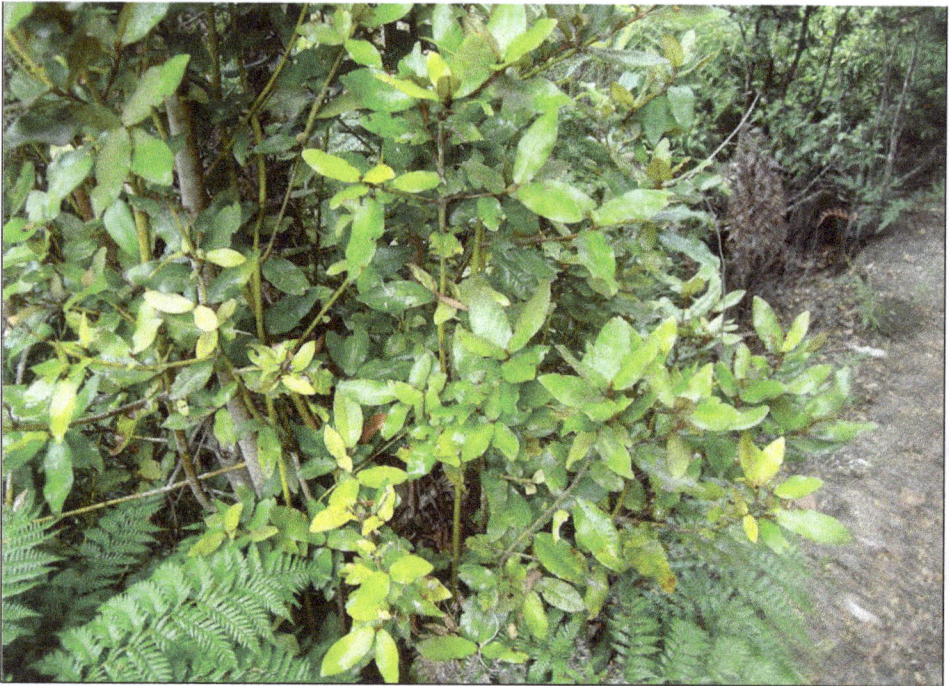

**Figure 2.38: Olivillo, Tique. *Aextoxicom punctatum* R. Et P. Aextoxicaceae.
@ Roberto Saavedra.**

Figure 2.39: Palo Santo. *Dasyphyllum diacanthoides* (Less.) Cabrera. Asteraceae. @ Roberto Saavedra.

**Espino negro.** *Rhaphithamnus spinosus* (A.L.Juss) Moldenke. Verbenaceae (Figure 2.40). Tree. This is an interesting tree that has been used for magical conceptions. Its thorns and the colour of the fruit are associated with a negative connotation. Thus, even the fruits are useful for medicine, they are not ingested or used for medicinal treatments.

**Figure 2.40: Espino negro. *Rhaphithamnus spinosus* (A.L.Juss) Moldenke. @ Roberto Saavedra.**

The previous plants mentioned with their classification and magical properties are part of the environmental area in southern Chile and the rest of their medicinal properties are not mentioned in this chapter. The area of study is considered a hot spot in biodiversity in the world and poses a huge amount of endemic species known by Williche communities. Both humans and plants are an indivisible connection that needs to be protected over time.

## 2.13 Conclusions

An acute information, insights, and alternative treatments of psychoactive plants species from indigenous knowledge are still to be learnt and documented from traditional healers regarding to their cosmovision, botanical knowledge, and healing practices that can contribute and offer new ways of treatments of mental ailments in particular and broadly to other ailments. Moreover, psychoactive plants require further studies that can cover many cultural aspects and cultural meaning of ailments such as mental ones. It is necessary to include their deep understanding of usage and treatments conducted inside an indigenous community and look at these aspects based on their own perspective; most of roles theses plants play into the spiritual practices of traditional healers and shamans; the interrelation and influence of various psychoactive plants species when uses as a primary treatment or in combination with other species that can originate a third and new treatment; the mythology under which these psychoactive plants species are based on the indigenous collective memory, and folklore from where these plants are included into a cultural array of indigenous groups from their own development. For that reason, more rigorous ethnobotanical studies are needed to cover more hidden details and interesting new thoughts that can capture much of the remaining medical wisdom and philosophy of healers and shamans and their particular knowledge accumulated from generation to generation. Further knowledge of this sacred and delicate space would help to understand a bit more about this phenomenon from the indigenous cultures, where a shaman is the real expert of a wide spectrum of states of consciousness and self-awareness which arises from the centre of the same human personality. As seen in this chapter for Mexican and Chilean plants, it can be said that both are related each other exisiting parallel concepts in usage, procedures, and purposes as well as use of species, but at the same time, both have their own development and identity which make them more interesting to investigate in the future.

This life motive from the ethnobotanical point of view may produce deep reflections and new understandings on how to find other ways to approach and solve real problems in modern society.

In regards to the concept of protection, many scientists from different areas of study know that healers and shamans provide significant intellectual guidance, so their rights should be protected as well as the communities and their forest areas in the region where a scientist work, and be able to establish ethnobiomedical reserves to ensure that medicinal plants will be available for local use too. Plant knowledge seems to be disappearing faster than the forests themselves if efforts are not taking into account. A shaman who is an expert of a wide spectrum of states of consciousness, realize self-awareness which arises from the centre of the human personality, so further knowledge of this sacred and delicate space would help to understand a bit more about this phenomenon from the indigenous cultures, and can also help to respect it and learn more from it.

Finally, psychoactive plants species are diverse either in Mexico and southern Chile; they are inserted into the cultural life of indigenous people and are their vehicles of healing for several cultural and spiritual ailments. Thus, further documentation of these species may provide information for conservation, sustainable utilization, and preservation of local traditional knowledge and cultural heritage.

## Acknowledgements

Special thanks to indigenous communities and, mainly Williche ones from southern Chile for their support and knowledge.

# Chapter 3

# Medicinal Plant for Maternal Healthcare

*James O. Fajemiroye*

## ABSTRACT

*Medicinal plants have played significant roles among rural women in the management of pre-pregnancy, pregnancy, birth, and postpartum medical needs. Generally, medicinal plants to reduce maternal mortality and ensure the birth of healthy children. These plants are applied to treat jaundice, malaria, stomach ache, constipation, fever and viral infections, mastitis, uterine prolapse, sleeping sickness, dysentery, sterility, vaginal infections, snake bite, epilepsy, abdominal pain, inflammation, asthma, bronchitis, fever and infants diarrhoea. They are used to ease labour during delivery, reduce menstrual bleeding and postpartum haemorrhage, alleviate menstrual, parturition and postpartum pain, increase or inhibit lactation and combat malnutrition among children and nursing mothers. They also serve as native soap to wash private part to clear fallopian tube blockage, as an expectorant and abortifacients among others. A search was done with the aid of Google Scholar, PubMed, Science Direct and books using wide varieties of relevant keywords to facilitate searching.*

## 3.1 Introduction

Pre-pregnancy, pregnancy, parturition and the *puerperium* are important phases of life and are not without risk to the mother and infant. Rural women are constantly subjected to medical traditions such as postpartum confinement, steam baths, food taboos and medicinal plants application. Many medicinal plants are traditionally prepared to cure and/or manage ailments associated with maternal health. The affordability, accessibility and the general assumption that being natural means safety have popularized the use of medicinal plants. People living in rural areas are always faced with poor healthcare facilities and lack of qualified professionals. Medicinal needs of women in rural areas during pregnancy, birth and postpartum period to cure pregnancy related ailments such as miscarriage, morning sickness, swelling of the legs abdominal pain, retained placenta, postpartum haemorrhage etcetera (Abdillahi and Van Staden, 2013) have made them resorted to the use of medicinal plant, that are readily available. Most rural women patronizes the services of traditional birth attendants and healers, who often use medicinal plants for maternal healthcare. The medicinal plants being used for the treatment of obstetric and gynaecological conditions, such as birth control, complications during pregnancy and childbirth, and problems associated with infertility have been reported (Borokini *et al.*, 2013; Abdillahi and Van Staden, 2013).

Medicinal plants contain numerous secondary metabolites (active or inactive) that often act synergistically to induce some effects in the biological system either beneficial or not. The findings on biological activities of plants in experimental animals seem to correlate with the folcloric use. It is important to note that plants are often combined for local preparation and, at times it is difficult to know which plant is responsible for what effects. Meanwhile, biological properties, safety or toxicity status of some of these species are yet to be fully investigated. For the purpose of this chapter, information on maternal uses, biological activities and toxicity studies of the most cited species below were sourced from Scopus, ISI Web of Science, Science Direct, SpringerLink, MEDLINE, PubMed Central, Google Scholar and text books.

## 3.2 Some Species with Local Indication and Preclinical Data

### *Achyranthes aspera* L.

*Achyranthes aspera*, known as prickly chaff flower belongs to the Family Amaranthaceae. It is commonly found in wastelands. The plant is used in treatment of asthma, bleeding, in facilitating delivery, boils, bronchitis, cold, cough, colic, debility, dropsy, dog bite, dysentery, ear complications, headache, leucoderma, pneumonia, renal complications, scorpion bite, snake bite and skin diseases, *etc.* (Jain, 1991). Traditional healers claim that addition of *A. aspera* would enhance the efficacy of any drug of plant origin. Extracts from roots of Achyranthes aspera have been reported to possess spermicidal activity in human and rat sperm, as studied by Paul *et al.* (2010). The aqueous methanolic extract of the whole plant have been shown to possess hypoglycaemic activity by Akhtar and Iqbal (1991). Alcoholic extract of the roots of *A. aspera*, exhibited anti-inflammatory activity in Wistar rats using carrageenan-induced paw edema method (Vijaya *et al.*, 2009). Gayathri *et al.* (2009) also reported antioxidant activity on leaves and roots. The ethanolic extract of *A. aspera* shows bronchoprotective effect in toluene diisocyanate induced occupational asthma in Wistar rats as reported by Goyal *et al.* (2007). The methanolic extract of the whole plant of *A. aspera* induced nephroprotective activity against lead acetate induced nephrotoxicity in male albino rats (Jayakumar *et al.*, 2009).

### *Acacia nilotica* (L.) Delile

*Acacia nilotica*, commonly known as acacia, Egyptian mimosa, Egyptian thorn, red thorn. Babool, babul, belongs to the Fabaceae family. The origin of the name Acacia means 'spiny' which is a typical feature of the species. *Acacia nilotica* is widespread in subtropical, tropical Africa, from Egypt to Mauritania and southwards to South Africa, Asia eastwards to Pakistan and India. It has been used since early Egyptian dynasties. Disocorides (the Greek philosopher, physician

## Table 3.1: Summary of Medicinal Plants Used in Women's Healthcare

| Scientific Name | Family | Preparation and Medicinal Use |
|---|---|---|
| *Achyrantus aspera* L. | Amaranthaceae | Root: powder is given once a day to women for easy delivery and seed powder with water thrice a day to cure the dysentery. |
| *Ageratum conyzoides* L. | Asteraceae | Decoction drink for Perineal healing; Retraction of the uterus |
| *Alpinia galanga* (L.) Willd. | Zingiberaceae | Roast-Eat; Decoction drink for Anaemia (dizziness, headache); Mild puerperal fever; Lactagogue |
| *Aloe barbedensis* Mill | Liliaceae | Leaf juice is used in dyspepsia, amenorrhoea, burns colic, skin diseases constipation, vitiated condition of vat, kapha pitta, abdominal tumors |
| *Amaranthus spinosa* L. | Amarantaceae | Cold infusion drink for postpartum varicella |
| *Amomum microcarpum* C. F. Liang and D. Fang | Zingiberaceae | Decoction wash for Infant fever, reduces temperature |
| *Annasas comosus* Linn. Merrill | Bromeliaceae | Fruit: The fruit as unripe and sour in taste it is directly used as uterine tonic.They are useful in vitiated condition of sexually transmitted disease, amenorrhoea. |
| *Anethum graveolence* Linn. | Apiaceae | The fruits are acrid, They are useful for amenorrhoea, dysmenorrahoea |
| *Artocarpus heterophyllus* Lam. | Moraceae | Cold apply, Decoction drink for Neonatal navel healing; Lactagogue |
| *Barringtonia longipes* Gagnep. | Lecythidacea | Roast-warm poultice or massage Lactagogue for Painful or hard breasts; Improve flow of milk |
| *Bambusa arundinaceae* (Retz.)willd | Poaceae | Grinded bud decoction after delivery to prevent pregnancy |
| *Bischofia javanica* Blume | Euphorbiaceae | Roast poultice or massage for painful or hard breasts; Improve flow of milk; Neonatal navel healing |
| *Blumea balsamifera* (L.) DC. | Asteraceae | Cold infusion drink for Infant diarrhoea |
| *Carica papaya* Linn. | Caricaceae | Fruits are used directly in severe constipation and urinary tract infection; fruit eaten for abortion directly; raw fruit as vegetables are used to improve lactation in feeding mother. Latex: The latex is used as galactagogue tonic. |
| *Bauhinia variegate.*. Linn | Ceasalpinaceae | Bud about 4-5 buds grinded and give with water twice a day to women for enhancing lactation of feeding mother |
| *Catharanthus roseus* Linn.G. Don | Apocynaceae | Leaves: An infusion of the leaves considered for the treatment of menorrhiage. |

*Contd...*

**Table 3.1–Contd...**

| Scientific Name | Family | Preparation and Medicinal Use |
|---|---|---|
| *Chromolaena odorata* (L.) R. King and H. Robinson | Asteraceae | Decoction drink for pre-menstruation pain; Perineal healing; Retraction of the uterus; Expel lochia |
| *Clitoria ternatea* Linn | Fabaceae | Blue pea flower can be used to arrest uterine haemorrage |
| *Commiphora wighitti* (Arn.) Bhandari | Buraceae | Leaves: The decoction of leaf used for white discharge |
| *Coriandram sativum* Linn. | | The leaves are acrid astringent,aromatic,anti-inflammatory, carminative constipation jaundice.The fruits are aromatic, emollient and cure stomachic, constipation, scrofula, helminthaiasis, intermittent fevers, gout and giddiness. |
| *Curcuma domestica* valeton | Zinziberaceae | Rhizome powder is used as Antiseptic, pain killer, blood purifier |
| *Cumminum cyminum* Linn. | Apiaceae | The fruit are acrid used to cure constipation, stomachic, diuretic, uterine and tonic is useful to haemorrhoids leucorrhoea, gonorrhea. |
| *Diospyros apiculata* Hiern | Ebenaceae | Fresh-eat Abortifacient |
| *Daucus carota* Linn. *var.sativa* | Apiaceae | Root juice to increase lactation in feeding mother.Seed decoction taken daily early in the morning in the empty stomach for five days to abort one month pregnancy.The extract of root drinking early in the morning to regulate menstrual disorder. |
| *Elephantopus scaber* Linn. | Asteraceae | The root decoction and leaves are given in dysuria, urethrorrhea, the root decoction is used for heamorrhoids and paste made out of leaves is very useful for skin disease. |
| *Ficus hispida* L.f. | Moraceae | Cold infusion drink for Neonatal rash after high fever |
| *Ficus religiosa* Linn | Moraceae | Fruit kheer one teaspoon thrice a day for a week to sessile women for bearing child |
| *Glochidion eriocarpum* Champ. | Euphorbiaceae | Decoction drink for anaemia (dizziness, headache); Mild puerperal fever; Abdominal pain |
| *Gonocaryum lobbianum* (Miers) Kurz | Icacinaceae | Roast-warm poutice or massage for Painful or hard breasts; Improve flow of milk |
| *Lawsonia inermis* L. | Lythraceae | Leaf juice mixed with water and sugar given as a remedy for heat control, cooling, headache relief |
| *Michelia champaca* Linn. | Mangnoliaceae | One flower bud consumed orally by women with water after menstruation for a week to avoid pregnancy, bark and root are used as contraceptives. |

Contd...

**Table 3.1–Contd...**

| Scientific Name | Family | Preparation and Medicinal Use |
|---|---|---|
| *Mimosa pudica* Linn | Mimoseae | Leaf juice after delivery to avoid pregnancy, leaf juice with black pepper twice a day to control fever after child birth. |
| *Moringa pterygosper Mw* (Retz)Roxb. | Moringaceae | Leaf decoction is used before delivery, for constipation; fruit and seed decoction to cure sexual weakness; powdered root and extract of root kept in vagina for abortion |
| *Ocimun sanctum* L. | Lamiaceae | Seeds demulcent used in genital urinary disorder; leaves decoction used for haemorrage |
| *Phoebe lanceolata* (Nees) Nees | Lauraceae | Warm poultice; massage for painful or hard breasts; Improve flow of milk |
| *Punica granatum* Linn | Punicaceae | Fresh flowers with sugar twice a day with water for 2-3 days to cure leucorrhoea. |
| *Riccinus communis* L. | Euphorbiaceae | Leaves are lactagogue and applied as poultice over the breast and taken internally in the form of juice. It is also used for contraceptives. Seed; the cotyledons of one seed taken early in empty stomach from 5th day of menstruation for 25 days to produce sterility |
| *Rowlfia serpentine* (L) Benth ex Kurz | Apocynaceae | Root: It is used as uterine contraction and promotes the expulsion of the fetus. |
| *Rubus cochinchinensis* Tratt. | Rosaceae | Roast decoction drink for Anaemia (dizziness, headache); Mild puerperal fever |
| *Rubus tonkinensis* F. Bolle | Rosaceae | Roast decoction drink for Anaemia (dizziness, headache); Mild puerperal fever |
| *Santalum album* L. | Santalaceae | Wood decoction with ginger is beneficial in haemorrhoids |
| *Saraca asoca* (Roxb.) de wilde | Ceasalpinaceae | The bark is bitter, used to cure;constipation. It is useful in menorrage the flower decoction is considered to be a uterine tonic |
| *Syzyium cumin* (Linn.) skeels | Myrtaceae | One teaspoon of dried powder of bark with cow milk twice a day for a week to cure leucorrhoea. |
| *Tacca chantrieri* André | Taccaceae | Decoction drink for Postpartum secondary bleeding; Perineal healing, Retraction of the uterus; Expel lochia; Abdominal pain |
| *Tamarindus indicus* L. | Fabaceae | Steambath; Decoction wash for Postpartum recovery, Varicella, Mild puerperal fever; Neonatal rash after high fever |
| *Tetracera scandens* (L.) Merr. | Dilleniaceae | Roast decoction drink; Fresh chew for Postpartum recovery, Postpartum secondary haemorrhage, Aperative, Infant oral candida |

*Contd...*

**Table 3.1–Contd...**

| Scientific Name | Family | Preparation and Medicinal Use |
|---|---|---|
| *Trevesia palmata* (Lindl.) Vis. | Araliaceae | Roast decoction drink for Postpartum recovery, Perineal healing, Retraction of the uterus; Expel lochia; Abdominal pain; Physical recovery; Lactagogue |
| *Trigonella foenumgrace* UM L. | Fabaceae | Seeds are bitter; used as galactogue, pain killer. |
| *Vetiveria zizanoides* (Linn) | Poaceae | Roots are used to cure constipation and hemorrhage, fruits boiled with jaggary once a day for 2-3 days to regulate menstruation |
| *Zingiber officinale* Ros. | Zizingiberaceae | Rhizome, raw as well as dried to cure constipation, cough cold, to cure morning sickness of pregnant ladies, pain killer |
| *Ziziphus funiculosa* Ham. | Rhamnaceae | Roast decoction drink for Postpartum, Anaemia (dizziness, headache), Mild puerperal fever; Physical recovery |
| *Ziziphus maurtiana* Lam. | Rhamnaceae | Bark powder for constipation; fruits for blood purification and used as tonic |

*Source:* Lamxay *et al.*, 2011; Manisha *et al.*, 2012.

and 'father of botany' c.40 to 90 A.D.) described 'akakia'in his Materia Medica - a preparation extracted from the leaves and fruit pods. The young leaves are used to manage venereal diseases (Acharya and Pokhrel, 2007) while its pods are being used as cure for postpartum wound healing. The dried bark is used to cure diarrhea (Agunu *et al.*, 2005) and female infertility (Soladoye *et al.*, 2014). The root and stem barks are used to cure sexually transmitted infections (Kambizi and Afolayan, 2001), fibroids and gonorrhea (Jiofack *et al.*, 2010), skin wounds (Gemedo-Dalle *et al.*, 2005), diarrhea, gargle, sore throat, eczema and chest complaints (Shah *et al.*, 2006). The fruits are used to treat malaria, furuncles and phlegmatic cough (Musa *et al.*, 2011) and gastric ulcers (Nadembega, *et al.*, 2011).

Preclinical studies of this species have shown its antimicrobial against *Escherichia coli, Staphylococcus aureus* and *Salmonella typhi, Candida albicans* and *Aspergillus niger* (Saini *et al.*, 2008; Malviya *et al.*, 2011). The breast milk enhancement property of *A. nilotica* has been reported in female rats. The leaves extracts of *A. nilotica* increased milk and prolactin release (Lompo-Ouedraogo *et al.*, 2004). Fruits methanolic extract showed activity against Newcastle Disease Virus (NDV) and Fowl fox Virus (Mohamed *et al.*, 2010). The antidiarrheal property of *A. nilotica* was confirmed in albino rats through a decrease in intestinal transit of charcoal in castor oil induced diarrheal (Sanni *et al.*, 2010). Antioxidant property of *A. nilotica* was supported by increase in antioxidant enzymes, inhibition of linoleic acid oxidation and DPPH radical scavenging activity (Malviya *et al.*, 2011; Sultana *et al.*, 2007). As serum parameters of hepatic and renal function, histopathological features in liver sections, fasting glucose and triglycerides in rats were unaltered by the administration of *A. nilotica*, thereby suggesting low toxicity (Mohan *et al.*, 2014; El-Hidayah *et al.*, 2011).

## *Acacia polyacantha* Willd

*Acacia polyacantha* commonly known as catechu tree, belongs to Fabaceae family. It is widely distributed in tropical Africa from the Gambia to Eritrea, Ethiopia, in the north, to the Transvaal in the south. It is being used for general wellbeing. The stem bark of *Acacia polyacantha* is used to treat fibroids, body pains and typhoid fever while juice obtained from squeezed leaf is being used for wound healing (Olajide *et al.*, 2013). The seeds are used as abortifacients and contraceptives (Adebisi and Alebiosu, 2014). The powdered root bark mixed with honey is used to treat cough and asthma while the stem bark is being used to treat jaundice (El-Kamali and El-Khalifa, 1999; Koné *et al.*, 2004), dysentery and gastric ulcer (Doka and Yagi, 2009), fibroids and gonococci (Jiofack *et al.*, 2010), mouth gargle (Juyal and Ghildiyal, 2013). The root extract is being used to treat sterility in women (Kamuhabwa *et al.*, 2000), body pain and body swelling (Madge, 1998). Stem bark is used to treat tuberculosis (Tabuti *et al.*, 2010).

Potent antioxidant activity of *A. polyacantha* has been reported using DPPH radical, superoxide anions and hydroxyl radical scavenging and DNA-protective properties (Guleria. *et al.*, 2011). The presence of phenolic and flavonoid compound in *A. polyacantha* has been partly attributed to its antioxidant property (Hazra *et al.*, 2010). The antibacterial (Lakshmi and Aravind kumar, 2011), antidiabetic (Okpanachi *et al.*, 2010), anticancer (Ghate *et al.*, 2014) of this species among others have been reported. *A. polyacantha* crude extract showed some signs of toxicity including convulsion, restlessness, abdominal writhing and death at the oral dose of 1600 mg/kg upwards (Okpanachi *et al.*, 2010).

## Acanthospermum hispidum DC

*Acanthospermum hispidum* commonly known as Starburr, bristly, starbur, goat's head, hispid starburr, belongs to Asteraceae family. It is naturalized in many scattered places in Eurasia, Africa and North America. It is widely used against diarrhea (Agunu *et al.*, 2005), jaundice, malaria, stomach ache, constipation, fever and viral infections (Adu *et al.*, 2011), sleeping sickness (Kamanzi *et al.*, 2004), dysentery (Novy, 1997), sterility, vaginal and parasitic infections (Koukouikila-Koussounda *et al.*, 2013), snake bite, epilepsy, abdominal pain (Chakraborty *et al.*, 2012), asthma, bronchitis, fevers; and as an expectorant and abortifacient (Lemonica and Alvarenga, 1994; Evani *et al.*, 2008).

The antibacterial (Adu *et al.*, 2011), anti-trypanasomal and anti-plasmodial (Kamanzi *et al.*, 2004; Koukouikila-Koussounda *et al.*, 2013), antidiarrheal (Agunu *et al.*, 2005), antiviral (Summerfield *et al.*, 1997), anti-tumor (Deepa and Rajendran, 2007) and anti-helminthic (Herekrishna *et al.*, 2010) of this species have been reported. The treatment of pregnant rat with the plant extract induced external malformations and foetuses with visceral anomalies (Lemonica and Alvarenga, 1994), liver necrosis, shrinkage of renal glomerular tufts, haemorrhage and congestion in the spleen, lungs and heart as well as mortality in mice (Bakhita and Adam, 1978) suggest toxic effect of this species.

## Albizia chevalieri Harms

*Albizia chevalieri* commonly known as flat crown, belongs to the Fabaceae family. It is found in southern Sahel and Northern Sudanian savannas from Senegal to Chad. It is being used as purgative, taenicidal (Aliyu *et al.*, 2009) and abortifacient (Adebisi and Alebiosu, 2014). Its leaf powder is mixed with butter to treat body swelling (Hussaini and Karatela, 1989). The leaf decoction is widely used to cure dysentery (Burkill, 1995); the bark for the treatment of fever, hypertension and stomach ache (Adelanwa and Tijjani, 2013). The leaves are mixed with *Alchorea coracifolia*, *Graminae spp* and *Pseudospondias microcarpa* to make vapour bath against a high fever (Terashima *et al.*, 1992).

The leaf extract showed antioxidant activity using DPPH radical scavenging model (Aliyu *et al.*, 2009). The leaf extract also showed strong antibacterial activity against *Staphylococcus aureus* and *Escherichia coli* (Kwaji and Japaru, 2015). In acute and sub-chronic toxicity tests of *A. chevalieri* leaf extract, no significant change was observed in weight, biochemical and histopathological parameters (Saidu *et al.*, 2007; Yusuf *et al.*, 2007).

## Allium sativum L.

*Allium sativum* is a strongly aromatic bulb commonly known as garlic. It belongs to Amaryllidaceae family. Garlic is native to Central Asia, northeastern Iran with widespread cultivation. It is being used to treat diabetes and hypertension

(Nwauzoma and Dappa, 2013), malaria, cleansing of blood and cold (Lifongo, Simoben, Ntie-Kang, Babiaka and Judson, 2014) and bone fracture (Pradesh, 2011), female infertility (Soladoye *et al.*, 2014), inflammatory diseases (Ogbole *et al.*, 2010). It is used as antiseptic, lenitive and antihelmintic (Maxia *et al.*, 2008). It is used for blood cleaning, counter-poison, charm and magic (Nedelcheva and Draganov, 2014). Its powder is mixed with yoghurt and applied on vagina to treat postpartum vaginal inflammation (Ali-Shtayeh *et al.*, 2015).

*In vitro* study of *A. sativum* extract showed inhibition activity against cytokines mediating inflammatory bowel diseases (Londhe *et al.*, 2011). Analgesic (Jayanthi and Jyoti, 2012), antioxidant (Park *et al.*, 2009), antidiabetic (Thomson *et al.*, 2007), anticancer (Yeung *et al.*, 2011), antiviral (Gebreyohannes and Gebreyohannes, 2013), antileishmanial (Mahmoudvand *et al.*, 2014), antinematode (Masamha *et al.*, 2010) and heavy metals intoxicant (Massadeh *et al.*, 2007) activities of this species have been reported. In toxicity study, no death and appreciable pathological lesion was observed in rabbits treated with up to 2200mg/kg aqueous extract (Mikail, 2010). However, change in liver enzymes was reported in Wistar rats following treatment with aqueous *A. sativum* (Sulaiman *et al.*, 2014).

## Anacardium occidentale L.

*Anacardium occidentale* commonly known as cashew, belongs to Anacardiaceae family. Originally, native to northeastern Brazil, the tree is now widely grown in tropical regions, India and Nigeria being major producers, in addition to Vietnam, the Ivory Coast, and Indonesia. It is being used to treat stomach ache during pregnancy, opportunistic infections (Mustapha, 2014), asthma (Sonibare and Gbile, 2008) and malaria (Ogie-Odia and Oluowo, 2009). It has been used as insect repellent (Innocent *et al.*, 2014). Juice from the fruit is being used for wound healing (Ayyanar and Ignacimuthu, 2009).

Some biological properties of *A. occidentale* include hypoglycemic (Fagbohun and Odufunwa, 2010), antimicrobial (Chabi *et al.*, 2014), antiulcer (Konan and Bacchi, 2007), antioxidant (Razali *et al.*, 2008), antiviral (Kudi and Myint, 1999). Although acute toxicity study did not show renal and hepato-biliary disfunctions (Konan *et al.*, 2007), liver and kidney toxicities were however, observed in mice (Tédong *et al.*, 2008).

## Anchomanes difformis (Blume) Engl.

*Anchomanes difformis* commonly known as forest anchomanes, belongs to the family Araceae. It is widely distributed in tropical Africa and used to treat fever and inflammation (Akah and Njike, 1990), hyperthermia (Idu *et al.*, 2010), cough (Okpo *et al.*, 2011), cure tooth ache (Abondo *et al.*, 1991) and river blindness (Gills, 1992). The plant rhizome is used as lactogenic (Raponda-Walker and Sillans, 1961).

The inhibition of formalin induced pain and egg albumin induced inflammation in rat, support antinociceptive and anti-inflammatory properties, respectively, of *A. difformis* (Adebayo *et al.*, 2014). Other studies have reported the antiinflammatory and analgesic (Akah and Njike, 1990; Akah and Nwambie, 1994), antioxidant (Abubakar *et al.*, 2013), antimicrobial (Eneojo *et al.*, 2011), gastroprotective (Okpo *et al.*, 2011), antipyretic (Gabriel *et al.*, 2013), anti-trypanosomal (Nwodo *et al.*, 2015) properties of this species. Renal toxicity following excessive consumption has been reported in rats (Ataman and Idu, 2015).

### Anisopus mannii N.E.Br.

*Anisopus mannii* known as 'Kashe zaki' (Hausa) meaning "destroying sweetness", belongs to the family Apocynaceae. It is widely used to treat diabetes (Abo *et al.*, 2008; Ezuruike and Prieto, 2013), high blood pressure and infants jaundice (Atawodi *et al.*, 2014).

Methanol extract of the aerial part of the plant showed analgesic and anti-inflammatory properties using acetic acid abdominal constriction in mice and carrageenan induced paw edema in rats, respectively (Musa *et al.*, 2009). In addition antioxidant (Aliyu *et al.*, 2010; Atawodi *et al.*, 2010), hypoglycemic (Sani *et al.*, 2009; Tijjani *et al.*, 2012; Zaruwa *et al.*, 2012), antimicrobial (Aliyu *et al.*, 2011; Musa *et al.*, 2015). Treatment with *Anisopus mannii* extract up to 400mg/kg for a period of 28 days did not show adverse effect on body weight, organ weight, serum electrolytes and most of liver function parameters in albino rats (Sani *et al.*, 2010). A high $LD_{50}$ value of 1250mg/kg and histopathological study suggest the safety of this extract in rats (Tijjani *et al.*, 2012).

### Artemisia annua L.

*Artemisia annua* commonly known as sweet annie, belongs to the family Asteraceae. It is an important medicinal plant in many ancient cultures especially Asia and Africa. It is widely used to manage fevers, remedy for summer heat, jaundice, wound healing, eye problems (Willcox, 2009; Meier zu Biesen, 2011; Ogwang *et al.*, 2012; Liu *et al.*, 2013; van der Kooy and Sullivan, 2013). It is also used to treat spasm, agitation, hemorrhage and diarrhea (Emadi *et al.*, 2010; Mirdeilami *et al.*, 2011).

Antimalarial activity of *A. annua* which was associated with artemisinin has been established (Liu *et al.*, 1992; Weathers *et al.*, 2014; Willcox *et al.*, 2004). Antihypertensive which was associated with decrease in heart beat rate and systolic blood pressure (Ben-nasr *et al.*, 2013), anti-inflammatory (Huang *et al.*, 1993; Melillo *et al.*, 2012), antimicrobial (Juteau *et al.*, 2002; Massiha *et al.*, 2013), anticancer (Cafferata *et al.*, 2010) antioxidant (Juteau *et al.*, 2002; Caver *et al.*, 2012) properties have been reported. Gastrointestinal symptoms such as nausea, vomiting, abdominal pain, and diarrhea (Willcox *et al.*, 2004) have been observed in 3.4 per cent out of five hundred and ninety patients that were treated with the *A. annua* extract. Toxicity studies in pregnant rats, however, showed post implantation losses, lower maternal progestagens and significant maternal testosterone (Boareto *et al.*, 2008).

### Baccharoides adoensis Sch.Bip. ex Walp.

*Baccharoides adoensis* belongs to the family Asteraceae. It is locally used against vomit, stomach ache (Deeni and Hussain, 1994), sexually transmitted infections, malaria, kidney and heart problems (Swamy *et al.*, 2013a), tuberculosis (Chitemerere and Mukanganyama, 2011), gastritis, abdominal pain, nausea, duodenal ulcers and for wound healing (Nergard *et al.*, 2004; Inngjerdingen *et al.*, 2012; Austarheim *et al.*, 2012).

Antimicrobial (Deeni and Hussain, 1994; Chitemerere and Mukanganyama, 2011; Swamy *et al.*, 2013b), antiulcer (Germano *et al.*, 1996; Austarheim *et al.*, 2012), immunomodulatory (Nergard *et al.*, 2004), anti-inflammatory (Ibrahim *et al.*, 2009) of *B. adoensis* have been reported. It was found to be moderately toxic ($LD_{50}$ = 471.2mg/kg) in experimental mice (Ibrahim *et al.*, 2009).

## *Caralluma dalzielii* N.E. Brown

*Caralluma dalzielii* belongs to the family Asclepiadaceae. It is widely used as antispasmodic and as analgesic remedy (De Leo *et al.*, 2005). It is also used to treat cardiovascular diseases, diabetes and blood cholesterol level (Tanko *et al.*, 2013), pain, fever, inflammation and to suppress appetite (Maheshu *et al.*, 2014), obesity, waist decrease, blood purification and to lower blood cholesterol level (Adnan *et al.*, 2014). Latex obtained from the stem is used for wound healing while the stem is eaten raw as tonic and to treat cardiac problems (Neuwinger, 2000).

Aqueous extract of *C. dalzielii* showed dose dependent anti-inflammatory and antinociceptive properties using acetic acid induced writhing test and sub plantar formalin induced nociception, respectively (Ugwah-Oguejiofor *et al.*, 2013). Other biological properties of *C. dalzielii* that have been reported include antimutagenic (Gowri and Chinnaswamy, 2011), antidiabetic and hepatoprotective (Latha *et al.*, 2014), antibacterial (Kulkarni *et al.*, 2012), anti-proliferative (Vajha and Chillara, 2014), hypolipidemic (Tanko *et al.*, 2013). In a toxicity evaluation test, no mortality, hematological, pathological and histopathological changes were observed when rats were treated with 2g/kg and 5g/kg for 14 days (Nutrients, 2014).

## *Citrus aurantifolia* (Christm.) Swingle

*Citrus aurantifolia* commonly known as lime tree, belongs to the family Rutaceae. Its mixture with *Mentha viridis* is used to treat nausea and vomiting during pregnancy (Acharyya and Sharma, 2004). It is also used with other species to treat morning sickness during pregnancy (Ticktin and Dalle, 2005). The decoction prepared from the leaves is used to cure headache, cold and fever (Muthu *et al.*, 2006; Ignacimuthu *et al.*, 2008), flu, wound and stomach ache (Wondimu *et al.*, 2007). Decoction prepared from the roots and leaves of this species alone or mixed with *Cassia alata* and *Dendrocnide simulans* is used to treat coughing with blood while leaves and roots are mixed with leaves and roots of other *Citrus* spp to cure fever (Grosvenor *et al.*, 1995; Amri and Kisangau, 2012). The fruit is rubbed on the scalp to cure dandruff and headache (Ong and Norzalina, 1999).

The essential oil from *C. aurantifolia* showed spasmolytic property on jejunum, aorta and uterus isolated from rabbit (Spadaro *et al.*, 2012). Antimalarial (Bapna *et al.*, 2014), antioxidant, antibacterial (Kaur and Mondal, 2014), anti-atherosclerosis, anti-cancer, anti-inflammatory and antimicrobial properties (Tripoli *et al.*, 2007), hypolipidaemic (Yaghmaie *et al.*, 2011) activities have been reported. The extracts of *C. aurantifolia* aerial part showed no sign of toxicity in brine shrimp lethality assay (Bhatt *et al.*, 2015). Although administration of *C. aurantifolia* root extract caused significant elevation of rats' liver enzymes without significant histopathological change (Chunlaratthanaphorn *et al.*, 2007). In another acute toxicity test, no lethality was observed when rats were treated with *C. aurantifolia* fruit juice extract up to 5000mg/kg, however, adverse effects were recorded on fertility indices (FSH, LH, sperm mobility, morphology and concentration, prolactin and testosterone) (Okon and Etim, 2014).

## *Clitoria ternatea* L.

*Clitoria ternatea* commonly known as Butterfly pea belongs to the family Fabaceae. It is native to the East Indies, China as well as Egypt. The plant is used to treat insect bites, skin diseases, asthma, burning sensation, ascites, inflammation, leucoderma, leprosy, hemicrania, amentia and pulmonary tuberculosis. It is commonly called "Shankpushpi" in the Sanskrit language where it is reported to be a good "Medhya" (brain tonic) (Daisy *et al.*, 2009).

The ethanolic extract of *Clitoria ternatea* Linn. Showed antioxidant activities in vitro (Parimaladevi *et al.*, 2003) by using DPPH free radical, Feric reducing power assay, super oxide dismutase and total poly phenols method. *Clitoria ternatea* has been evaluated for its medicinal properties and shows promising effects as having antioxidant, antidiabetic and hepatoprotective activities. Chronic administration of plant extracts (100mg/kg) for 14 days reduces the blood glucose level of the diabetes induced animals (Wistar Albino rats) as compared to diabetic control group (Gunjam *et al.*, 2010), thereby confirming antidiabetic property of this species.

Ethanolic leaf extract of *Clitoria ternatea* was evaluated for prophylactic and therapeutic hepatoprotective activity against carbon tetrachloride induced hepatic damage (Somania *et al.*, 2011). Hepatoprotective effect of EECT was evident in prophylactic and therapeutic groups at doses of 200 and 400 mg/kg. Histopathology of liver ascertained the effect of EECT and carbon tetrachloride on cytoarchitecture of the liver. The administration of carbon tetrachloride in animals showed severe centrilobular necrosis, fatty changes, vacuolization and ballooning degeneration indicating severe damage of liver cytoarchitecture (Somania *et al.*, 2011).

## *Euphorbia balsamifera* Aiton, Hort.

*Euphorbia balsamifera* commonly known as balsam spurge, belongs to the family Euphorbiaceae. It is used traditionally as insect repellant (Suleiman *et al.*, 2014). The decoction prepared from the branches is used to alleviate whooping cough (Nadembega *et al.*, 2011), hemorrhoids, intestinal worms and skin diseases (Mathieu and Meissa, 2011), wound, toothache (Ellena *et al.*, 2012a). The root decoction is used to wash chronic wounds (Inngjerdingen *et al.*, 2004), treat skin infections and gonorrhea (Adedapo *et al.*, 2004).

Both roots, stems and leaves of *E. balsamifera* showed good antimicrobial activity (Kamba and Hassan, 2010). Insect repellence activity of *E. balsamifera* against bean beetles (*Callosobruchus maculatus*) and mosquitoes (*Anopheles gambiae*) have been reported in experimental models (Suleiman *et al.*, 2014). Toxicity study in rats revealed that *E. balsamifera* caused lycocytosis with significant increase in albumin, alanine aminotransferase (ALT) aspartate aminotransferase (AST) and reduction in globulin (Adedapo *et al.*, 2004).

## *Evolvulus alsinoides* (L.) L.

*Evolvulus alsinoides* commonly known as dwarf morning glory, belongs to the family Convolvulaceae. It is widely distributed in tropical countries. It is an important spiritual and medicinal plant in Ayurvedic system of medicine being used for general wellbeing. It is principally used as charm for love and favour (Burkill, 1985), pain killer (Abubakar *et al.*, 2007) and asthma (Indhumol *et al.*, 2013). It is used to manage insanity, epilepsy, nervous debility, loss of memory (Siripurapu *et al.*, 2005; Austin, 2008); cures infectious diseases such as syphilis, diarrhea, dysentery and leucorrhea (Jain *et al.*, 2008; Alagesaboopathi, 2009). It is used against bowel irregularities also as vermifuge and febrifuge (Singh, 2008).

Its adaptogenic and anti-amnesic properties were proved with reduction in stress-induced perturbation, improved peripheral stress markers and scopolamine induced dementia (Siripurapu *et al.*, 2005). The immunomodulation property of *E. alsinoides* was confirmed through reduction in synovial hyperplasia and nitric oxide synthase (NOS) in adjuvant induced arthritic rat model (Ganju *et al.*, 2003). Other biological activities of *E. alsinoides* include gastroprotective (Hewageegana *et al.*, 2006), antidiabetic (Gomathi *et al.*, 2013), hypolipidemic (Iyer and Patil, 2011), anti-inflammatory, antipyretic and antidiarrheal (Lekshmi and Reddy, 2011), anxiolytic (Nahata *et al.*, 2009), anti-dyskinesia (Rahman *et al.*, 2010), blood purification (Hebbani *et al.*, 2014), memory restoration (Naidu *et al.*, 2013); hepatoprotective (Thatipelli and Yellu, 2014) and hair growth promotion (Amrita *et al.*, 2012). No toxicity and lethality were observed in albino mice treated with alcoholic extract of *E. alsinoides* (Agarwal and Dey, 1977). In a chronic toxicity test, haematological parameters such as serum glutamic oxaloacetic transaminase, serum glutamic pyruvic transaminase, urea, cholesterol and fasting blood glucose were slightly reduced while serum triglyceride was significantly reduced (Hewageegana *et al.*, 2006).

## *Guiera senegalensis* J.F. Gmel

*Guiera senegalensis* commonly known as Moshi medicine, belongs to the family Combretaceae. The plant is used to cure malaria, diarrhea, dysentery, venereal diseases and microbial infections (Lifongo *et al.*, 2014), cough, syphilis, leprosy, impotence and gastroenteritis (Adamu *et al.*, 2005), and neutralize snake venom (Sallau *et al.*, 2005). The fruits are used to manage uterine prolapse, abdominal pain and cough. The galls are used to treat various infants' illnesses, wound, nightmares, skin diseases, high blood pressure, tuberculosis and cough (Nadembega *et al.*, 2011). The root bark extract is used to treat hypertension (Jiofack *et al.*, 2010), stems are used to cure abscess, toothache (Tapsoba and Deschamps, 2006), diarrhea (Koné and Atindehou, 2008). The root bark is used to manage infectious diseases (Magassouba *et al.*, 2007) while the leaves are used to treat malaria (M. S. Traore *et al.*, 2013) and jaundice (EL-Kamali, 2009). The twigs and shoots are used to manage epilepsy and convulsion (Pedersen *et al.*, 2009).

The extract of this species alleviated experimentally induced diarrhea (Williams *et al.*, 2009). An hydroacetic extract of the galls showed significant antioxidant activity (Sombie *et al.*, 2011). The chloroform root extract of *G. senegalensis* showed antimalarial activity against *Plasmodium falciparum* (Ancolio *et al.*, 2002). The antiviral activity of *G. senegalensis* leaves extract against poliovirus, human herpes virus, equine herpes virus and astrovirus has been reported (Kudi and Myint, 1999). Significant reduction in the mortality of albino mice following intraperitoneal administration of reconstituted venom from *Echi carinatus* and *Naja nigrocollis* incubated with *G. senegalensis* leaves extract confirmed its snake venom detoxifying activity (Abubakar *et al.*, 2000). Sedative property of *G. senegalensis* extract was proven by its ability to reduce spontaneous motor activity in mice and increase the duration of pentobarbital sleeping time (Amos *et al.*, 2001). Bako *et al.* (2014) showed that 500 and 1000mg/kg doses of the methanol extract had no effect on liver function indices but significantly increased the level of potassium and bicarbonates. It was concluded that the extract could lead to disturbance in acid-base balance as a result of hyperkalemia brought about by erythrocyte haemolytic effect. Elrahman *et al.* (2008) reported that Wistar albino rats treated with 1000 and 500mg/kg/day methanol leaf extract died within one week of the experiments with endotheliotoxicity, hepatonephropathy and hyperplasia. In contrary, the treatment of Wistar rats with aqueous extract of *G. senegalensis* leaves for six months did not show any toxicity based on renal and hepatic functions, biochemical and organs histopathological analyses (Diouf *et al.*, 1999).

### *Ipomoea asarifolia* (Desr.) Roem. And Schult

*Ipomoea asarifolia* commonly known as morning glory plant, belongs to the family Convolvulaceae. Its leaves are used for the treatment of gastrointestinal disorders and diabetes (Atawodi and Onaolapo, 2010), malaria (Nda-Umar *et al.*, 2014), urinary problems in pregnancy, hemorrhage and as abortifacient (Farida *et al.*, 2012). The leaf juice is used to cure skin infections (Xavier *et al.*, 2015), itching, dermatitis and eczema (Aleixo *et al.*, 2014).

The hepatoprotective activity of this species has been shown in carbon tetrachloride-induced hepatotoxicity model (Farida *et al.*, 2012). Egg albumin-induced inflammation and formalin-induced pain were significantly reduced after treatment with methanolic leaves extract of *I. asarifolia* (Jegede *et al.*, 2009). *Ipomoea asarifolia* is identified as poisonous plant (Agaie *et al.*, 2007) and (Medeiros *et al.*, 2003). Carvalho de Lucena *et al.* (2014) found that the tremorgenic compound of *I. asarifolia* are eliminated in milk and the milk can intoxicate nursing lambs. It was found that the plant affects motor coordination (Riet-correa *et al.*, 2014). Nine out of ten goats treated with 5-37g/kg *I. asarifolia* leaves daily showed tremorgenic syndrome and death of two with no observable lesion at necropsies (Medeiros *et al.*, 2003).

### *Jatropha curcas* L.

*Jatropha curcas* commonly known as Barbados nut, belongs to the family Euphorbiaceae. It is a drought resistant tropical plant native to America and Caribbean. It is a multipurpose plant with many attributes and potentials and it is specially recognized for the production of biodiesel from its seed. All parts of the plants are used to cure various diseases and conditions including HIV/AIDS (Agbogidi *et al.*, 2013), irregular mensuration, herpes, rectal enema, whitlow and convulsion (Nwauzoma and Dappa, 2013). The juice from *J. curcas* leaves is mixed with *Pterocarpus osun* (camwood's) bark and native soap to wash private part three times for one month to clear fallopian tube blockage (Borokini *et al.*, 2013). It is used to treat dental complaints and constipation (Dada *et al.*, 2014). The leaves are also mixed with *Azadarichta indica* (neem) and *Carica papaya* (papaya) leaves to treat malaria (Olatokun and Ayanbode, 2009). Leaves decoction is used as lactagogue, rubefacient, anthelmintic and as medication of urinary ailments (Iwu, 1993). The decoction prepared by mixing the leaves with leaves, roots and fruits of *Xylopia ethiopica* (African pepper) is used to cure edemas, cough, drepanocytosia, haemorrhoids, jaundice and liver problems (Neuwinger, 1996). The leaves are used to promote lactation and as medicine against dysentery, colic, amenorrhea, oligomenorrhoea, hypertension and headache (Jain and Srivastava, 2005; Parveen and Shikha, 2007; Nath and Choudhury, 2010), while the twigs are used against toothache inflammation and bleeding of the tooth gum (Sharma *et al.*, 2009). The leaves are also used as galactagogue and for homeostasis (Abdelgadir and Van Staden, 2013).

The anti-inflammatory activity using carrageenan induced rat paw edema among other models has been reported (Nayak and Patel, 2010; Oskoueian *et al.*, 2011; Othman *et al.*, 2015). Leaves methanol extract inhibited egg albumin induced edema in Wistar albino rat (Uche and Aprioku, 2008). Antioxidant activity of this species using ferric ion reducing, DPPH radical, nitric oxide (NO), superoxide anions (SO) and hydrogen peroxide ($H_2O_2$) scavenging assays has been reported (Diwani *et al.*, 2009; Igbinosa *et al.*, 2011; Oskoueian *et al.*, 2011). *J. curcas* leaves extract showed analgesic activity in hot plate and acetic acid induced writhing models in mice and rats (Yusuf and Maxwell, 2010), antiviral (Wender *et al.*, 2008), hepatoprotective (Balaji *et al.*, 2009), anticancer (Theoduloz *et al.*, 2009; Liu *et al.*, 2012), antidiabetic (Jayakumar *et al.*, 2010; Mishra *et al.*, 2013) and wound healing (Esimone *et al.*, 2008; Balangcod and Balangcod, 2011) activities. The toxicity of *J. curcas* are mostly atributed to the seeds and fruits of the plant (Li *et al.*, 2010; Singh *et al.*, 2010; Shah and Sanmukhani, 2010). In another study, *J. curcas* ameliorate chloroform induced hepatotoxicity by lowering the levels of alanine amino transferase (ALT) aspertate amino tranferase (AST) and alkaline mino transferase (ALP) in albino rats (Okechukwu *et al.*, 2015).

## Lafoensia pacari A. St. Hil.

*Lafoensia pacari* belonging to Lythraceae family is commonly found in South and Central America (Cabral and Pasa, 2009). This species has been used to treat pneumonia (Bueno *et al.*, 2005) and ulcers (Guarim Neto, 2006; Souza and Felfili, 2006). The leaf infusion of *L. pacari* has been used as diaphoretic (Mendonça *et al.*, 2006). The crude extract from inner bark of *L. pacari* showed a promising antioxidant (Solon *et al.*, 2000; Burque *et al.*, 2015) and anti-inflammatory (Roger *et al.*, 2008) activities. The anticancer (Marcondes *et al.*, 2014), antiviral (Muller *et al.*, 2007) and antimicrobial activities of *L. Pacari* have also been reported. These biological activities have been associated to the presence of secondary metabolites such as tannins, flavonoids, steroids, saponins, phenolic compounds and others in *L. pacari* (Solon *et al.*, 2000).

## Lannea acida A. Rich.

*Lannea acida*, commonly known as African grape tree, belongs to the family Anacardiaceae. It is among the plants used to treat ulcer and diarrhea (Etuk *et al.*, 2009; Oluranti *et al.*, 2012), stomach pain (Obata and Aigbokhan, 2012), antimalarial and anti-convulsive (Asase *et al.*, 2005; Ziblim *et al.*, 2013), gonorrhea and rheumatism (Koné *et al.*, 2004). It showed antioxidant activity (Ouattara *et al.*, 2011), antibacterial activity against *E. coli* (Aboaba *et al.*, 2006).

The hydro-alcoholic bark extract showed inhibition of *Mycobacterium tuberculosis* H37R$_v$ proliferation and immunostimulating property (Ouattara *et al.*, 2011). The bark extract of *L. acida* showed antidiarrheal activity in Sprague-Dawley rat (Etuk *et al.*, 2009). Its pro-fertility property was also reported (Ahmed *et al.*, 2010). It showed no toxicity on HeLa cervix adenocarcinoma cell line (Sowemimo *et al.*, 2009).

## Leptadenia hastata (Pers.) Decne

*Leptadenia hastata*, commonly found in open area or climbs on bushes, belongs to the family Asclepiadaceae. It is widely consumed by pregnant women (Umaru *et al.*, 2014). It is widely used as food and for the treatment of hypertension and catarrh (Dambatta, 2011), urinary schistosomiasis and onchocercosis (Bah *et al.*, 2006; Togola *et al.*, 2008), sexual potency, skin diseases treatment, scabies and wound healing (Tamboura *et al.*, 2005; Betti *et al.*, 2011; Belem *et al.*, 2007). It is used as galactagogue and purgative (Thomas, 2012). Other biological activities of the plant include antimicrobial (Aliero and Wara, 2009; Anywar *et al.*, 2014), anti-androgenic (Bayala *et al.*, 2011; Bayala *et al.*, 2012), anti-inflammatory, analgesic and wound healing (Kuribara *et al.*, 2001; Sarkiyayi *et al.*, 2015), hypoglycemic (Bello *et al.*, 2011a,b) and anti-spermatogenic (Bayala *et al.*, 2011).

Evaluation of leaf extract of *L. hastata* reduced biochemical parameters (fasting blood sugar (FBS), total cholesterol (TC), triglycerides (TG) and low density lipoprotein, urea creatinine and alanine aminotransferase) in pregnant rat (Umaru *et al.*, 2014). A high LD quotient value of 0.78 was observed after 48-72 hours of mice treatment (Tamboura *et al.*, 2005). No cytotoxicity was observed in human lymphoblastoid cells after 6 hours of incubation (Aquino *et al.*, 1996). Degenerated sites of implantation and foetal resorbtion were induced in pregnant albino rats (Garba *et al.*, 2013).

## Mangifera indica L.

*Mangifera indica*, commonly known as mango, belongs to the family Anacardiaceae. It is used to treat all sorts of fever and female infertility (Ajaiyeoba *et al.*, 2003; Odugbemi *et al.*, 2007; Soladoye *et al.*, 2014), malaria, diarrhea, diabetes, hypertension, hemorrhoid, insomnia, insanity, asthma, cough; and as astringent, emmenagogue (Olowokudejo *et al.*, 2008) and insect repellent (Mavundza *et al.*, 2011). It is used to treat HIV/AIDS opportunistic infections (Mugisha *et al.*, 2014), diabetes treatment (Ocvirk *et al.*, 2013) and digestive problems (Volpato *et al.*, 2009).

The bark aqueous extract of *M. indica.* significantly inhibit carrageenan induced inflammation in rat (Oluwole and Esume, 2015). The aqueous leaf extract of *M. indica* also reduced writhing response (Islam *et al.*, 2010). In another study, *M. indica* extract inhibited the induction of $PGE_2$, tumor necrosis factor alpha factor (TNF-α) serum level and ear edema induced by arachidonic acid (Garrido *et al.*, 2004). Various parts of the plant showed considerable antioxidant activity using DPPH radical scavenging and linoleic inhibition capacity (Sultana *et al.*, 2012). Other biological properties of *M. indica* include antidiabetic (Aderibigbe *et al.*, 2001), hepatoprotective (Rodeiro *et al.*, 2008), anticancer (Prasad *et al.*, 2008; Joona *et al.*, 2013), immunomodulatory (Makare *et al.*, 2001) antimicrobial (Singh *et al.*, 2010; El-Gied *et al.*, 2015) and intoxicant (Lakshmi *et al.*, 2011). No histopathological change was observed in the liver of *M. indica* aqueous bark extract in Wistar rats (Oyewo *et al.*, 2012). However, the administration *of M. indica* aqueous leaf extract in female Sprague-dawley rat reduced weight gain, serum follicle stimulating hormone (FSH) and induced estrous cycling disruption (Awobajo *et al.*, 2013).

## *Mitragyna inermis* (Willd) Kuntze

*Mitragya inermis*, commonly known as false abura, belongs to Rubiaceae family. It is widely used to treat stomach ache and intestinal disorders (Sy *et al.*, 2004), diabetes, ulcers, pile, dysentery, bone pain (Adoum *et al.*, 2012) and hypertension (Igoli *et al.*, 2005). The decoction prepared from its stem bark is used for the treatment of chickenpox and wart (Balde *et al.*, 2015), jaundice (Karou *et al.*, 2011). The leaves and twigs are combined with whole *Indigofera pulchra* to treat malaria (Alex *et al.*, 2005).

The species showed hypoglycemic activity in alloxan induced diabetic albino rats (Adoum *et al.*, 2012). *M. inermis* showed inhibition of *Staphylococcus aureus* (Zongo *et al.*, 2009), *Bacillus subtilis*, *Pseudomonas syringae* and *Cladosporium herbarum* (Asase *et al.*, 2008). Both aqueous and ethanolic extracts of *M. inermis* increased the onset of pentylenetetrazol and strychnine induced clonic convulsion in albino rats, suggesting its anticonvulsant property (Timothy *et al.*, 2014). The antihypertensive activity of *M. inermis* has been reported (Ouédraogo *et al.*, 2004). *M. inermis* inhibited ileal basal tonus and submaximal contraction thereby confirming its use in treating intestinal disorders (Sy *et al.*, 2004). No evidence of mutagenic or genetotoxic activity was observed with an alkaloid rich extract of *M. inermis* (Traore *et al.*, 2000). The albino rats treated with aqueous extract of *M. inermis* showed significant increase in the levels of aspartate aminotransferase and alanine aminotransferase were observed (Timothy *et al.*, 2015).

## *Moringa oleifera* Lam.

*Moringa oleifera*, commonly known as drumstick tree or horse radish tree, belongs to the family Moringaceae. *M. oleifera* is distributed in many tropic and sub tropic countries and regarded as important plant because of its remarkable medicinal value. It is referred to as miracle tree because of its wider therapeutic applications. The leaves are used to treat fever, sore throats, bronchitis, eye and ear infections (Morton, 1991), skin, digestive, respiratory and joint ailments (Cáceres *et al.*, 1991), "blood booster" especially for women and children (Otitoju *et al.*, 2014). It is also used to combat malnutrition among children and nursing mothers (Fahey, 2005;

Dhakar *et al.*, 2011), for pregnant women to prevent maternal anemia and low birth weight (Iskandar *et al.*, 2015). In a study, 50 per cent increase in hemoglobin content in pregnant women and absence of low birth weight were observed following the intake of *M. oleifera* extract (Iskandar *et al.*, 2015).

Investigations on the influence of *M. oleifera* leaves aqueous extract on haematological indices in Wistar rats revealed that hemoglobin count, packed cell volume and white blood cells were significantly increased while the amount of red blood cells was significantly reduced, thereby suggesting a possible hematotoxicity (Otitoju *et al.*, 2014). In another study, ethanolic leaves extract of *M. oleifera* was found to mitigate AlCl$_3$ induced anemia in albino rats as it significantly increased the RBC, mean corpuscular hemoglobin concentration (MCHC), hematocrit and hemoglobin (Osman *et al.*, 2012). Idohou-Dossou *et al.* (2011) however, report that although hemoglobin concentration was significantly increased in lactating women placed on *Moringa* leaves powder. Ethanol extract of *M. oleifera* was also found to decrease serum bilirubin, increase albumin, total protein, hemoglobin, hematocrit; (Ujah *et al.*, 2013). In another study, no significant difference in sperm quality, haemotological and biochemical (liver enzymes, urea and creatine) parameters between aqueous leaf extract treated Wistar albino mice and control (Awodele *et al.*, 2012). Scherl *et al.* (2010) reviewed the toxicity studies on *M. oleifera* in both human and animal studies and concluded that many preparations including aqueous extract appeared to be "exceedingly safe at the doses and in the amount commonly used."

### *Ozoroa insignis* R. Fernandes

*Ozoroa insignis*, commonly known as Eastern Cape resin tree, belongs to the family Anacardiaceae. It is used to treat Pneumonia and opportunistic infections (Kisangau *et al.*, 2007; Otieno *et al.*, 2011), schistisomiasis (Ndamba *et al.*, 1994), malaria (Asase *et al.*, 2005; Nadembega *et al.*, 2011), tooth ache and snake bite (Muthee *et al.*, 2011). Various studies have reported the biological properties of *O. insignis* including anticancer (Rea *et al.*, 2003), antihelmintic (Mølgaard *et al.*, 2001), antidiarrheal (Sibandze *et al.*, 2010), antioxidant, antimicrobial and antilypoxygenase (Ahmed *et al.*, 2014), antibacterial (Ishak *et al.*, 2013). The plant was found to be toxic against brine shrimps larvae (He *et al.*, 2002; Moshi *et al.*, 2004).

### *Saraca asoca* (Roxb.) W. J. de Wilde

*Saraca asoca*, known as Ashoka (which is a Sanskrit term meaning one 'without sorrow or grief') belongs to the family Caesalpiniaceae. The use of *S. asoca* dried bark, root and flowers have been used to treat many different gynaecological disorders (uterine abnormalities, menorrhagia (excessive menstrual bleeding), ammenorrhea, painful periods, endometrosis and disorders of the menstrual cycle (Mollik *et al.*, 2010; Saha *et al.*, 2012; Bhandary *et al.*, 1995; Kumar *et al.*, 1980; Middelkoop and Labadie, 1985). It has been used as uterotonic, anticancer, antibacterial, antihelminthic, antiulcer, antioxidant, refrigerant, hypolipidemic, hypoglycaemic, anti-inflammatory, analgesic, febrifuge, skin infection curative agent along with many other health benefits. A decoction which is prepared from the bark of *S.asoca* in water in combination with other herbs such as *Terminalia chebula* and *Coriandrum sativum* is are used by women suffering from menorrhagia (Saha *et al.*, 2012; Begum *et al.,* 2014) and to stop abnormal vaginal discharges (Saha *et al.,* 2012). The bark of this species is used to treat excessive salivation; flowers as a cardiac and brain tonic; while fruits are mentioned as contraceptive in nature(Biswas and Debnath, 1972).

Experiments have revealed the ability of bark aqueous extract to stimulate and relax the intestinal muscle, prolong uterine contractions and also as a uterine sedative (Pradhan *et al.,* 2009; 21). The antibacterial properties of methanolic, ethnolic, acetone and aqueous extracts of bark, dried flower buds and leaves of *S. asoca* have been reported (Satpal *et al.,* 2015). Anticancer (Yadav *et al.,* 2015), Anti-inflammatory (Saha *et al.,* 2012), antiarthritic, Antiulcer, Antioxidant, antidiabetic and hypolipidemic, analgesic (Satpal *et al.,* 2015), dermatoprotective (Pradhan *et al.,* 2009) have also been reported.

## *Sclerocarya birrea* (A. Rich.) Hochst

*Sclerocarya birrea,* commonly known as Marula, belongs to the family Anacardiaceae. It is used to treat malaria, wound healing, diabetes and enhance breast milk (Gouwakinnou *et al.,* 2011), cough, female infertility (Senkoro *et al.,* 2014a), rheumatism, dysentery, diarrhea (Peter, 2013), and diabetes (Dimo *et al.,* 2007). Biological activities of *S. birrea* include anticonvulsant (Ojewole, 2006), antioxidant, anthelmintic, anti-plasmodial, antibacterial, anti-inflammatory, hypoglycemic (Ojowele *et al.,* 2010). No toxicity signs were observed in mice (Ojowele *et al.,* 2010) and brine shrimps (McGaw and Eloff, 2008). Significant concentration dependent reduction in cell viability was observed in LLC-PK1 following 48 and 72 hours of treatment with stem bark extract (Gondwe *et al.,* 2008).

# Chapter 4

# Ethics and Procedural Steps in Basic Researches

*James O. Fajemiroye and Gustavo R. Pedrino*

## ABSTRACT

*This chapter has an objective of providing basic guidance on the ethics, principles and procedures in basic research. Millions of bench work are in progress on daily basis with the objective of new discovery and advancement of science. Meanwhile, the emerging challenges in the field of research and dissemination of experimental findings call for the regulation of scholarly activities towards the promotion of good science that are acceptable to scientific and non-scientific community. The aspect of laboratory safety, health issues, laboratory animals and procedural considerations towards the attainment of set objectives in vivo testing are explored.*

## 4.1 Laboratory Safety and Health Measures

As laboratory safety and health measures are prerequisites to basic researches, every prospective trainee needs to undergo safety training to prevent injury and cases of laboratory accidents before, during and after research. An effective hazard identification and risk assessment could prevent or reduce human exposure. Knowledgeable health and safety specialists are involved in the development of procedures to manage risks or potential experimental hazards such as biologic agents (*e.g.*, infectious agents or toxins), chemical agents (*e.g.*, carcinogens and mutagens), radiation (*e.g.*, radionuclides, X-rays, lasers), and physical hazards (*e.g.*, needles and syringes). The risks associated with unusual experimental conditions such as handling of chemicals, animal bites, exposure to allergens, chemical cleaning agents, wet floors, cage washers and other equipment, lifting, ladder use, and zoonoses that are inherent in or intrinsic to animal use should be identified and evaluated in terms of severity, the exposure intensity, duration, and frequency, susceptibility of researcher and the history of occupational illness and injury in the particular workplace.

However, in case of accidents or injury, experimenter should always report to superior laboratory attendant, advisor or any other designated authority. Emergency cases should be handled in a manner that safeguard life. Experiment can be put on hold, while burners or other energy-producing and energy-consuming devices are turned off as the experimenter evacuate the area as soon as the alarm goes off. Experimenter should always use appropriate apron, safety glasses, goggles, shields, gloves, masks, face shields, head covers, coats, or 'overalls', shoes or shoe covers in the laboratories. Safety glasses should be worn at all times while in the laboratory. Googles must be worn when working with corrosive materials (acids/ bases). Gloves are first line of defence to prevent contact with chemicals. Always wear the appropriate gloves for the material you are using. Rubber aprons should be used against strong acids and bases. Laboratory coats are intended to protect clothing not you. Never bring lab coats or aprons home.

Chemical fume hoods are used to control exposures toxic substances. Learn how to use a fume hood and know how to adjust air flow. If you have any concern(s) about safety and health issues in the laboratory, always contact your supervisor or any other designated authority on safety related issues. Before you use any chemicals you must be familiar with characteristics of the particular chemical. Pay careful attention and use appropriate protective equipment when working with novel/ synthetic compounds extracts, diagnostic specimens or compounds of undetermined toxicity. Always follow established laboratory procedures and techniques while using these materials. Treat all unknown compounds and investigational materials as toxic. Never use a standard household refrigerator or freezer for storage of flammable or reactive chemicals. Never eat, drink, smoke, or use cosmetics in areas that use or handle hazardous materials. Hazardous chemicals such as carcinogens, perchloric acid (in 72 per cent or greater concentration), aromatic amines, ethers, bromine, organic halides, carbon disulfide, cyanides, picric acid and radioisotopes pose a danger to human health or environment, if improperly handled. Common classes of hazardous chemicals are ignitable, corrosive, reactive and toxic. Never transport hazardous chemicals in personal vehicles.

Hazardous materials are defined as solid, semi-solid, liquid, or gas that, "because of its quantity, concentration, or physical, chemical, or infectious characteristics, it may cause, or significantly contribute to, an increase in mortality or an increase in serious irreversible, or incapacitating reversible, illness or pose a substantial present or potential hazard to human health and the environment when improperly treated, stored, transported, or disposed of, or otherwise managed." The disposal of hazardous, medical and radiological wastes should take health and safety into consideration. Never clean up a chemical spill unless you are familiar with the materials. If you do not know the hazards involved, or if you do not have the necessary supplies or protective equipment. Researcher should fully understand the hazardous characteristics of the specific chemicals being used.

A manufacturer's material safety data sheet remains the best source of information on hazards associated with the material one is working with as it provides protective equipment information. The warning signs on containers, minimum quantity of material that is toxic or hazardous, the specific routes of entry (skin or eye, by ingestion, by inhalation, through injection), the type(s) of hazard(s) (corrosive, explosive, flammable, reactive, sensitizer, toxic), the types of injury the material can cause (acute toxicity, chronic toxicity, carcinogen, mutagen, and teratogen), the symptoms of overexposure and the target organs that may be involved, the physical characteristics of the material (solid, liquid, and gas), vapor density, vapor pressure, flammability, chemical compatibility and incompatibilities, personal protective equipment for safe handling. Personal protection from unnecessary exposures could include working in a fume hood; never work alone when handling hazardous materials; clean working area, using of eye protection, wearing of a clean laboratory coat, use of ideal glove type for the materials you are handling, washing of hands before leaving work area (never attempt to eat, drink or use bathroom without washing hand), ensure identification through labelling of every container that holds both harmless and hazardous materials, keep storage containers tightly closed, organize containers in an easily identifiable and compatible manner. Any questions or concerns relating to health and safety for handling or disposal of chemicals should be directed to appropriate designated authority. Waste must be placed in good and compatible containers with label indicating full description of its content. Designate a single location that is accessible and do not obstruct traffic for the storage of hazardous waste be it a solid, semi-solid, liquid, or gas. Prompt attention and response should be paid to hazardous materials spill or accident. Special attention must be paid to a time-limited chemicals, (Ethers, etc.). because of the potential for peroxide formation. The date of delivery of time-limited chemicals should be recorded for safety use. Chemical that becomes unstable due to long storage be safely disposed. Refrigerators intended for laboratory use, should not be used for the storage of food products. Transportation of cylinders with cart should be done with the cap attached. In the laboratory, high pressure gas cylinders must be immobilized in an upright position and used with a regulator valve. Corrosive chemicals should always be transported in unbreakable safety carriers. Proper disposal of hazardous materials (glassware, empty bottles, etc.) and avoidance of indiscriminate disposal of hazardous chemical substance into the sanitary sewage system, atmosphere, or normal waste bin is a responsibility of every laboratory workers. Good orientation on emergency procedures in the laboratory is essential to safety.

As a general rule, safety depends on trained personnel who rigorously follow safe practices. Personnel at risk should be provided with clearly defined procedures and, in specific situations, personal protective equipment to safely conduct their duties, understand the hazards involved, and be proficient in implementing the required safeguards. When selecting specific safeguards for animal experimentation with hazardous agents, careful attention should be given to procedures for animal care and housing, storage and distribution of the agents, dose preparation and administration, body fluid and tissue handling, waste and carcass disposal, items that might be used temporarily and removed from the site (*e.g.*, written records, experimental devices, sample vials), and personal protection. Protective clothing and equipment should not be worn beyond the boundary of the animal facility

(DHHS, 2009). If appropriate, personnel should shower when they leave the animal care, procedure, or dose preparation areas. The use of good personal hygiene often reduces the possibility of occupational injury and cross contamination. Safety measure should also include protection of personnel against threats, harassment, assault, arson, and vandalism posed by those that are not comfortable with laboratory animals research (Please see more on the synopsis of chemical safety regulations University of Wisconsin-Madison Safety Department and Chemical Safety Manual, University of Mississippi).

## 4.2 Laboratory Animals

It remains an irony that biomedical scientists are in a collision course with the society which they intend to serve. Unethical conduct, poor policy and conflicting cultural value system are the bane of societal alienation to animal experimentation. Animal's welfares have been a topical issue among scientist, policy maker and society for years. Irrespective of individual's background and objective, global nature of science imposes an ethical obligation on researchers to promote societal interests with the best and acceptable animal experimentation. Despite the existing pocket of agitations against animal experimentation, the use of diverse species of animals in research is growing on a daily basis. Ethical concerns and the search for alternatives to animal experimentation has remain a huge challenge. Ideally, the care and overall welfare of all the animals should be a paramount responsibility of those that produce and use them for research and teaching. The care of experimental animals is a *sine qua non* to best practices of biomedical research. The important nature of animal care has led to the emergence of various institutions, policies, regulations, international and local laws around the world with the aim of enforcing humane care, high ethical standards, ensuring purposeful and responsible animal experimentation.

The quality of research, welfare of both animals and humans is an inevitable responsibility of any researcher in their pursuit of scientific knowledge. The expectation of a better understanding, new knowledge that could improve human and/or animal well-being (Perry, 2007) remained one of the finest justification for animal experimentation. The all-important principle of 3 Rs (replacement - avoid using animals; refinement - alter experimental procedures to minimize or eliminate pain and maximize welfare; and reduction - use of fewer animals for a comparable and logical conclusion) by Russell and Burch (1959) has been widely accepted as a practical strategy for a rational use of animals. The policies, procedures, standards, organizational structure, staffing, facilities, and practices put into place by an institution to achieve the humane care and use of animals in the laboratory and throughout the institution are critical to animal care and use.

According to the U.S. government, principles for utilization and care of vertebrate animals used in research supports the consideration of non-animal testing or less invasive procedures, experimental procedures that are relevant to the good of society, use of appropriate number of quality species, experimentation with minimal or without distress and pain, adequate use of sedative, analgesia, and anesthesia, humane endpoints, adequate veterinary care, appropriate transportation and execution of experimentation by qualified staff. Animal care personnel and other member of research teams (principal investigators, study directors, research technicians, postdoctoral fellows, students, and visiting scientists) should be trained to gain knowledge and expertise for effective utilization the specific animal procedures proposed and the species. A trained researcher is expected to have a detailed experimental protocol that describes in clear and concise terms the rationale behind the use of a specific species, number and group of animals as well as experimental endpoint. Researchers should identify, explain, and include in the animal use protocol a study endpoint that is both humane and scientifically sound. An inclusion of a humane endpoint (point at which pain or distress in an experimental animal is prevented, terminated, or relieved) the use of fewer animals and non-animal testing could indicate the consciousness of researcher towards animal care. It is equally essential to ensure the exposure of experimental animals to an appropriate environments that are well suited for the species or strains of animals being proposed for experiment. One should consider animals welfare (physical and mental needs), growth and reproduction. An experimental protocol should describe the nature of device being used to restrict normal movement of animal and number of animal in the cage prior to or after drug administration, collection of samples, and evaluation. Restraint devices should be appropriately designed to attain research goals while minimizing or preventing discomfort, pain, distress, and potential injury to the animal and the researcher. The animals are sensitive to the changes in temperature, humidity, illumination, noise, vibration, ventilation, etc.

Exposure to wide or extreme fluctuations in temperature and humidity may result in behavioural, physiologic, and morphologic changes that could have a negative impact on the animal's welfare and overall outcome of research. An healthy body temperature of the animals should be maintained within normal circadian variation. Animals should be housed within temperature. The thermoneutral zone – TNZ (bounded by the lower - LCTs and upper critical temperatures – UCTs) varies among species; Animal could adopt wide range of behavioural activities (foraging, digging, burrowing, nest building and huddling for resting and sleeping) to thermoregulate and control their microclimate at varying temperature (cold or warm). The quantity of heat being produced by lighting animal rooms should be set to avoid heat stress or skin burn in extreme condition. Animals should be provided with adaptable humidity ranges. The acceptable range of relative humidity is considered to be 30 per cent to 70 per cent for most mammalian species. Meanwhile, housing design, construction material, nesting material, number of animals, age, sex, species and size of the animals could contribute to inter and intra cage variation in temperature and humidity. Identification cards with the record of animal's source, strain or stock, sex, names and contact information for the responsible investigator(s), arrival date, birth date, protocol number or genotype information (when applicable) among others should be kept for individual animals.

The quality and quantity of illumination can also elicit physiological, morphological, and behavioural changes of various animals. Potential photostressors could emanate from inappropriate photoperiod, photointensity, and spectral quality of the light (Stoskopf 1983). Light intensity and wavelength as well as the duration of the animal's current and prior exposure to light, and the animal's pigmentation, circadian rhythm, body temperature, hormonal status, age, species, sex, and stock or strain can affect animals' needs for light (Duncan and O'Steen, 1985; Wax, 1977). Lighting should be evenly diffused throughout an animal holding area and provide sufficient illumination for the animals' well-being while permitting good housekeeping practices, adequate animal inspection including for the bottom-most cages in racks, and safe working conditions for personnel. Light intensity may differ as much as 80 fold in transparent cages from the top to the bottom of a rack, and differences up to 20-fold have been recorded within a cage. Hence, light intensity decreases with the square of the distance from its source. The practice of cage shifting relative to the light source or providing conditions that facilitate animal's behaviour like nesting, bedding or tunneling can reduce inappropriate light exposure.

Forced ventilation of enclosures, the type of bedding material (that absorbs urine and faces to facilitate cleaning and sanitation) and frequency of its changes (Besch, 1980) are also critical to environmental hygiene. The primary purpose of ventilation is to provide appropriate air quality and a stable environment. Specifically, ventilation provides an adequate oxygen supply; dissipates thermal loads caused by the animals, personnel, lights, and equipment; dilutes gaseous and particulate contaminants including allergens and airborne pathogens; adjusts the moisture content and temperature of room air. The type and location of supply air diffusers and exhaust registers in relation to the number, arrangement, location, and type of primary and secondary enclosures affect how well the microenvironments are ventilated. Static isolation caging (without forced ventilation), such as that used in some types of rodent housing, restricts ventilation. A good contact bedding, reduction in the number of animal per cage, frequent sanitation and cage change could engender better maintenance of static isolation caging.

As the activities of the researchers and the social interaction of the animals could produce noise and vibration, careful operation by researchers and purposeful design of animal facilities that resist sound and vibration could mitigate negative impact of these environmental-induced discomfort. Many species can hear sound frequencies inaudible to humans. Wider ranges of sound decibels could produce auditory and non-auditory effects. Animal facility should be located far away from human activities that constitutes noise and vibration. The degree of noise or vibration being produced by different animal species needs to be put into consideration when allocating animal facilities to animals. As it may be unsafe to interfere with the innate ability of the animals to make noise, external factors could be manipulated to accommodate noisy animals. The intensity, frequency, rapidity of onset, duration, and vibration potential should be evaluated. Proper record of sound and the hearing range, noise exposure history, and sound effect susceptibility of the species, stock, or strain should be recorded. Habituating animals to routine experimental procedures/condition should be encouraged in order to reduce the captivity/environment/novelty/people induced shock or stress. Odour emanating from the cage or researchers could interfere with the animal behaviour. This is the more reason why researcher needs to do away with perfume or cosmetics that could induce olfactory stimulation.

Animal care through all phases of growth and reproduction demands access to food and water *ad libitum*. In some cases, regimental regulation of food and fluid intake may be required for the attainment of research protocols. The regulation process may entail fasting, scheduled access to food or fluid sources in which the total volume of food or fluid consumed is strictly monitored and controlled (NRC 2003b). The planning and execution of a food and fluid restriction to achieve prior scientific objective should ideally not compromise the animals's welfare. The development of animal protocols that involve the use of food or fluid regulation need assessement of the degree of regulation, potential adverse consequences of regulation, and health implication (Morton, 2000; NRC, 2003b). Food or fluid restriction can be by the species, strain, gender, age of the animals, thermoregulatory demand; type of housing; time of feeding, nutritive value, and fiber content of the diet, prior experimental manipulation. The nutritional needs of the animals should be not be compromised by food and fluid restrictions. Daily records of food and fluid consumption, hydration status, body weights and any behavioural alteration should be noted (NRC, 2003b). Feed vendors should provide periodic data from laboratory-based feed analyses for critical nutrients. The user should know the date of manufacture and other factors that affect the food's shelf life. Stale food or food transported and stored inappropriately can become deficient in nutrients. Animal colony managers should be judicious when purchasing, transporting, storing, and handling food to minimize the introduction of diseases, parasites, potential disease vectors, and chemical contaminants in animal colonies. Purchasers are encouraged to consider manufacturers' and suppliers' procedures and practices (*e.g.*, storage, vermin control, and handling) for protecting and ensuring diet quality. In addition, animals should have access to potable and uncontaminated drinking water. Periodic monitoring for pH, hardness, and microbial or chemical contamination may be necessary to ensure that water quality is acceptable, particularly for use in studies in which normal components of water in a given locality can influence the results. Water can be treated or purified to minimize or eliminate contamination when protocols require highly purified water. The selection of water treatments should be carefully considered because many forms of water treatment have the potential to cause physiologic alterations, reduction in water consumption, changes in microflora, or effects on experimental results (NRC, 1996). Watering devices, such as drinking tubes and automated water delivery systems, should be checked frequently to ensure appropriate maintenance, cleanliness, and operation.

The pharmaceutical-grade chemicals and other substances should be used when available, for all animal-related procedures in order to prevent toxic or unwanted side effects on the experimental animals. Attention should be paid to the grade, purity, sterility, pH, pyrogenicity, osmolality, stability, site and route of administration, formulation, compatibility, and pharmacokinetics of the chemical or substance to be administered (NIH, 2008). The use of non-pharmaceutical-grade chemicals or substances should be described and justified.

Planning and design of animal housing systems requires professional knowledge and judgment and depends on the types of animals used, the nature of the hazards in question, the limitations, or capabilities of the facilities. Experimental animals should be housed so that possibly contaminated food and bedding, faces, and urine can be handled in a controlled manner. Social animals should be housed in stable pairs or groups of compatible individuals unless they must be housed alone for experimental reasons or because of social incompatibility. Important considerations for determining space include the age and sex of the animals, the number of animals to be cohoused and the duration of the accommodation, the use for which the animals are intended (*e.g.*, production vs. experimentation), and any special needs they may have. Breeding animals will require more space, particularly if neonatal animals will be raised together with their mother or as a breeding group until weaning age. Group composition is critical and numerous species specific factors such as age, behavioural repertoire, sex, natural social organization, breeding requirements, and health status should be taken into consideration when forming a group. In addition, due to conformational differences of animals within groups, more space or height may be required to meet the animals' physical and behavioural needs. Therefore, determination of the appropriate cage size is not based on body weight alone, and professional judgment is paramount in making such determinations. Structural adjustments are frequently required for social housing (*e.g.*, perches, visual barriers, refuges), and important resources (*e.g.*, food, water, and shelter) should be provided in such a way that they cannot be monopolized by dominant animals.

Sanitation is critical to the health and well-being of animals. This can be ensured through bedding change, cleaning, and disinfection. Cleaning removes excessive amounts of excrement, dirt, and debris, and disinfection reduces or eliminates unacceptable concentrations of microorganisms. The frequency and intensity of cleaning and disinfection should depend on what is necessary to provide a healthy environment for an animal. Methods and frequencies of sanitation will vary with many factors, including the normal physiologic and behavioural characteristics of the animals; the type, physical characteristics, and size of the enclosure; the type, number, size, age, and reproductive status of the animals; the use and type of bedding materials; temperature and relative humidity. Waste disposal Conventional, biologic, and hazardous waste should be removed and disposed of regularly and safely. There are several options for effective waste disposal. Contracts with licensed commercial waste disposal firms usually provide some assurance of regulatory compliance and safety. On-site incineration should comply with all federal, state, and local regulations. Hazardous wastes that are toxic, carcinogenic, flammable,

**Figure 4.1: Male *Swiss Albino* Mice (weight 25-30 g, 4 weeks)
Distributed in Home Cage Prior to Ventilated Cabinet Exposure
(Temperature–22°C; Relative humidity–36).**

corrosive, reactive, or otherwise unstable should be placed in properly labeled containers and disposed of as recommended by occupational health and safety specialists. In some circumstances, these wastes can be consolidated or blended. Sharps and glass should be disposed of in a manner that will prevent injury to waste handlers.

Animal facilities may be subject to unexpected conditions that result in the catastrophic failure of critical systems or significant personnel absenteeism, or other unexpected events that severely compromise ongoing animal care and well-being (ILAR, 2010). Facilities must therefore have a disaster plan. The plan should define the actions necessary to prevent animal pain, distress, and deaths due to loss of systems such as those that control ventilation, cooling, heating, or provision of potable water. Animals that cannot be relocated or protected from the consequences of the disaster must be humanely euthanized. The disaster plan should identify essential personnel who should be trained in advance in its implementation. Efforts should be taken to ensure personnel safety and provide access to essential personnel during or immediately after a disaster.

## 4.3 Procedural Considerations Prior to *in vivo* Testing

Psychotropic plants are chemical storehouses of active principles with potential therapeutic benefits. In order to increase the chances of identifying bioactive plant extracts, selection strategies such as use of traditional knowledge as clues, literature search for knowledge of major constituents, focus on plants at unique growth habitats, selection of plants rich in specific classes of constituents that have specific biological activity and knowledge of a common mode of action can be adopted. Although some of the empirical use of medicinal plants can be misleading, at times, they could serve as a clue to facilitate identification, collection and extraction of botanicals. However, once plants are identified and selected by indication, botanicals need to be extracted into test samples prior to the isolation of active principles as described by Williamson *et al.* (1996). Some investigators use alcoholic or methanol extraction as their initial crude extraction method (Li *et al.*, 2005) prior to further fractionation. Traditional methods used in the extraction of botanical samples for phytochemistry investigations often use sequential solvent extraction methods starting with less polar solvents, such as hexane, petroleum ether, ethyl acetate, and chloroform, followed by solvents with increasing polarity such as ethanol and water.

For instance if researcher is interested in recovering the major or all components for bioactivity screening, the use of aqueous alcohol (*e.g.*, 70 per cent when solvent penetration ability is strong) may be a simple and effective way of extracting polar and non-polar components at once. Other extraction methods like Pressurized liquid extraction, Microwave-assisted extraction, Supercritical Fluid Extraction, ultrasound-assisted extraction, steam distillation, and superheated-water extraction have been employed in the extraction of compounds from plant materials, Huie, 2002; Ong, 2004). With botanical materials/test substance other bench works are important prior to in vivo assays. The stability of extract components could be influenced by pH or the form of the extract. Components in the extracts are presumably much more stable structurally in powder forms than in solutions. Solution stability study may be carried out to monitor bioactivity (Liu, 2014).

The robustness and reproducibility of an assay depend on the type of signal measured (biochemical, physiological, behavioural, etc.), instrumentation and procedure employed. Biological system (Mouse, Rat, Hamster, Rabbit, Dog, Other rodent), number of animals to be used, study types (acute, subchronic or chronic toxicity tests), sex, strain, number of groups and animals in each groups, dose, route of administration (oral, intravenous, intramuscular, intraperitoneal, subcutaneous, inhalation, topical, Other) should be determined (UNICEF/UNDP/World Bank/WHO, 2012; Turner *et al.*, 2011).

In the absence of previous record on the dose of substance to be tested, a superior dose which could produce some evidence of toxicity, an intermediate dose and an inferior dose which does not influence animals could be tested; Selection of the appropriate dose form for administration to animals is both a science and an art. Dose of substance could make it beneficial or poison. Dose range and interval could vary depending on the efficacy and potency of the substance under investigation. Sometimes dose-response effect could be possible through a sufficient interval among the extrapolated doses. A shorter-term studies of doses may provide an insight for longer-term investigations.

As techniques, being employed in substance administration could interfere with behavioural response and the scientific validity of experimental results, substance preparation and delivery to animals in active form without contaminants should be carefully considered (Turner *et al.*, 2011). When the drug is given as a solution or suspension, the vehicle must be selected carefully, to avoid introducing confounding effects in the study. The amount of solution to be prepared depends on the species, study type, duration, and the vehicle used. Ionizing agents, co-solvents, surfactants, suspension dosing, emulsions among other methodologies can be used to optimize substance preparation. Attention should be paid to physical and chemical properties to prevent irritation and secondary effects on animals.

Test substance can be administered through enteral and parenteral routes. Enteral (oral) administration involves substance passage through the gastrointestinal tract (esophagus, stomach, and small and large intestines). Gastric gavage of test substance can be achieved by its administration in pellets, powdery form, aqueous solutions or through voluntary consumption in water bottles affixed to the cage.

The choice of parenteral routes (intramuscular, intravenous, intra-arterial, intra-thecal, intrapleural, intraperitoneal, subcutaneous, etc) could be determined by the duration of administration (a single instance or a session of several hours, repeated or intermittent administrations for days, weeks or months). The equipment for substance administration includes acute and chronic restraint devices, hypodermic needles, superficial or surgically implanted catheters and vascular access ports, gastric tubes, gavage needles, and infusion–delivery pumps. Minimizing pain, discomfort, infection, thrombosis, and loss of patency, particulates-induced embolization are primary considerations in vascular access. In addition, hygiene, sterility, routine device maintenance, accuracy of dosing, safety of animals and research personnel should be ensured. Appropriate use of anesthesia, analgesia, aseptic technique, and anticoagulant solutions are key to success (Turner *et al.*, 2011). Gentle cleansing with an antibacterial scrub or a disinfectant of the clipped puncture site should be carried out (Huerkamp, 2002). As drug administration sometimes involves some form of isolation and restraint, attention should be paid to animals visual, auditory, and olfactory contact with conspecifics and other environmental factors. Initial habituation to administration procedures and restraint devices could minimize stress and engender humane treatment (Cinelli *et al.*, 2007).

Test substances can be administered to animals as solutions (a homogeneous fluid mixture of 2 or more substances that cannot readily separate-out) or suspensions (a heterogeneous fluid mixture that contains solid particles that are dispersed throughout the liquid phase and that may sediment). The excipients (additives used to convert pharmacologically active compounds into pharmaceutical dosage forms suitable for administration) which could act as solvents, binders, surfactants, stabilizers, emulsifiers, lubricants, sweeteners, etc. should be carefully selected. Excipients could be referred to as vehicle when it is used as a medium in which the test substance is dissolved or suspended for administration (Rowe *et al.*, 2009). Reports on specific vehicles, their properties, applications, and toxicities can be reviewed in the literature (Gad *et al.*, 2006; Li and Zhao, 2007; Lipinski *et al.*, 2001; Neervannan, 2006; Rowe *et al.*, 2009; Smolinske, 1992; Weiner and Kotkoskie, 2000).

Contrary to some of the erroneous assumption, no vehicle is truly innocuous and without biologic effects (sterile water and 0.9 per cent saline, can cause deleterious unintended consequences, such as fluid overload, when given intravenously, or unintended diuresis (Hickling and Smith, 2000). Surfactants like polysorbates - Tween 80 and polyoxyl castor oil - Cremophor EL could act as wetting agents, reduce or eliminate drug precipitation, decrease drug degradation, enhance solubility control pH, modulate drug release, facilitate drug uptake (Li *et al.*, 1999; Li and Zhao, 2007). Techniques such as warming, shaking, sonication, and vortex mixing could also facilitate substance dissolution. Although gentle heating may assist with solubilization of substances, heating processes can cause precipitation or physical degradation of test substance (Hickling and Smith, 2000). Heat induced compound saturation in different physical phases may lead to particle separation when cooled. It is important that the compound remain suspended during dosing to ensure accuracy, and this goal can be achieved by use of a low speed stir plate with magnetic stir bars (Turner *et al.*, 2011). If the compound comes out of suspension, then the quantity of administered compound can become a significant variable in the study. The compound could be in solution at lower dose and a mix of solution and suspension at high dose (Turner *et al.*, 2011; Neervannan, 2006). High viscosity, particle aggregation, and caking attributes of high dose suspensions often have physical stability issues that could constitute difficulties in administration and heterogeneous drug distribution (Neervannan, 2006).

Soybean oil, lecithin, glycerine, corn and coconut could be used as emulsifying agents (Gad *et al.*, 2006). Emulsions are 2-phase systems that consist of oil and water with particles stabilized by surfactants in interfacial phases (Turner *et al.*, 2011). Ionizing agents (citric acid, sodium bicarbonate, maleic acid, and bases that ionize at physiologic pH, that is, from pH 2 to 9) are aqueous solutions can also provide a buffered pH for optimal dissolution of compounds and maintain drug solubility at the buffer pH (Turner *et al.*, 2011). Co-solvents like ethanol, propylene glycol, polyethylene glycol, and glycerine, DMSO are used to enhance the solubility of poorly water-soluble compounds (McCarthy, 2000; Thackaberry *et al.*, 2010). Cyclic oligosaccharides (Cyclodextrins) with varying sugar moieties could form water-soluble ligand–drug complexes with hydrophobic molecules to enhance solubility. Pilot in vitro solubility testing of a substance in the proposed vehicle should be used before any novel substance is administered to an animal, to reduce unnecessary animal use, ensure accuracy of dose calculations, and minimize waste of the novel substance or drug being administered (Higuchi and Connors, 1965).

The homogeneity, stability, formulation shelf-life, physical and chemical nature of vehicles (pH, osmolality, and viscosity) are important, particularly for long-term nonclinical safety studies. Viscosity is particularly important when considering the ease of delivering the material through a small-gauge intravenous or gavage needle. The pH of the dosing solution or suspension is a critical factor that could influence pharmacokinetic and induces tissue injury (Kostewicz *et al.*, 2002). Administration of solutions or suspensions with a highly acidic or alkaline pH can cause diarrhea, vomiting, tissue ulceration or necrosis at the site of catheter entrance or exit or gastric ulceration and pain (Turner *et al.*, 2011).

The use of reference drugs which will be compared with negative control or vehicle treated group is important to validate the models been used. It is advisable that the reference drug should be dissolved in the same vehicle with the same treatment regime as the test substance. Labels that contain the substance name, strength, and amount; preparation and expiration dates; and vehicle should be sticked to vials, tubes among others (Rich, 2004). Doses, volumes, and frequency of administration should be calculated, verified, and documented (Turner *et al.*, 2011).

A clear workspace should be used for different doses of substance preparation and an organized system of administrations to minimize dosing errors. Animal should be arranged in different groups with a cage card label and treated in a predetermined order (vehicle treated animal (control), test substance and references from lowest dose to the highest dose). Animals can also be marked on the tail, head or body for identification. Whenever possible, purchased or synthesized materials for administration should be of pharmaceutical grade, and free of impurities, including foreign bodies, concomitant substances, and signal impurities, such as chemical intermediates, isomers, and by-products, organic volatile agents, residual solvents, and toxic impurities (Ahuja, 2007; ILAR, 2010).

Some basic considerations in *in vivo* models including illumination pattern, humidity, ambient temperature, noise level, odour, stress and sex of animals, contextual manipulation of animal, treatment pattern, issues of false positive effects will be approached in **chapter 4.** Preclinical evaluation of these plants are vital to ensure scientific validation of traditional use and quality control. Series of well-validated and accurate models may lead to new pharmacological treatments that could be beneficial in clinical practice. Preliminary evaluations, Biological activity - Relevance of dose and route of administration, preliminary evaluations, determination of dose, determination of route of administration. The fact that medicinal plants have been used by man for several years' seems to make in vitro investigation a non-issue.

# Chapter 5

# From Field to Herbarium: Plant Collection and Preservation Techniques

*James O. Fajemiroye, Aderoju Adeleke Adesiyan*
*and Heleno Dias Ferreira*

## ABSTRACT

*A collection of pressed and dried plant specimens which is known as herbarium are common practice among botanists. The techniques for the collection and preservation of plants are prerequisite to scientific studies of medicinal plants. The collection of plant material is as important as reporting informations that are relevant to the sample material. Hence, advance preparation and a lot of effort are required towards quality plant collection. Accurate identification of plants in the field and effective sampling are essential for permanent record for future reference. This chapter highlighted some processes that are involved in the collection and preparation of herbarium specimen.*

## 5.1 Introduction

The plants collected in the field are properly prepared and incorporated in a herbarium of a given institution or research centre. Herbarium is where dry samples of plant species are kept according to a classification system. The word herbarium replaced such earlier terms as hortus siccus (dry garden) or hortus hyemalis (winter garden). Botanist Luca Ghini (1490 - 1556) is considered to have been the first person to dry plants under pressure, mount them on paper, and thus preserve them as a permanent record (Bridson and Forman 1992). All kind of plant samples including ferns, bryophytes, gymnosperms, angiosperm, fungi, algae among others can be found in herbarium. The herbarium is an important centre of reference where researchers, taxonomists among other experts do consultation for plant identification. The specimens that are deposited in the herbarium often allow accurate identification of plants, provide a permanent record for a species occurring at a particular time and place, provide information on plant distribution, document the introduction and spread of invasive weeds over time, serve as reference point for the application of the scientific names and as vouchers for seed collections, biodiscovery, toxicological and biochemical analyses.

The collections that are identified and deposited in the herbarium are used by professionals from different fields including pharmacy, biochemistry, genetics, ecology, etc. Herbarium could serve as a centre for the exchange of botanical materials with other educational institutions, teach courses, train students, and receive researchers. The notes in the label are important sources of data for researchers on genetic relationships, morphology, anatomy, geographical distribution, flowering, fruiting, soil, ecology, taxonomic and phylogenetic data regarding the voucher specimen. A voucher herbarium specimen could be regarded as a compressed plant sample deposited for future reference. It supports research work and may be examined to verify the identity of the specific plant used in a study. Hence, a voucher specimen must be deposited in a recognized herbarium committed to long-term maintenance.

## 5.2 Field Works

The field collections can be faster or longer, depending on the objectives. Historically, the first collectors of plants in the world for scientific study were botanists in Europe. They collected not only in Europe but also in other continents, such as Asia, Africa and the New World, etc. The collected materials were taken to Europe and incorporated into the major European herbaria, for example, the herbarium at Kew and the Museum of Natural History in Paris. The collection of plants began in the 16 th century. Later, J.P.Tourefort (ca 1700, France) used the term herbarium for plants (Bridson and Forman 1999). Plant collections are essentials for taxonomic researches because they serve as voucher specimens. They also help to identify the family, genus and species. So, a herbarium is basically a storehouse of botanical specimens, which are arranged in the sequence of an accepted classification system, and available for reference or other scientific study (Maden, 2004). Plant collection techniques involves the use of instruments such as pruning shears, a pair of secateurs, scrapers, diggers, presses, rope to tie presses, cardboard and old newspapers to assist the drying of the collected plant samples, a field notebook for notes of data from specimens and the environment region where the plantsn are found, pencil, pen, eraser, caliper, *tape to take action*, plastic collecting bags, tape, paper bags, labels, cords, hand lens, GPS to record the geographic coordinates and altitudes, long-sleeved shirt and long trousers to protect against sunlight and leggings for snakes attack protection, a jumper and water-proof raincoat to keep the cold and rain off, a first-aid kit, water, food and a trip plan outlining your intended destination(s) and expected time of return left with someone who will raise help if necessary.

Researchers should avoid the collection of plants that have been damaged by insect or under deterioration. The specimens should be representative of the population, but should include the range of variation of the plants. Roots, bulbs, and other underground parts should be carefully dug up, and the soil removed with care. At least three branches of fertile samples (about 20 to 30 cm in length) of leaves, flowers and/or fruits should be collected from shrubs and tree. Herbaceous plants up to 40 cm height and grasses should be collected at the distal ends. The fruit must be collected and tagged according to the respective sample numbers. A fleshy fruit should be placed in glasses containing 70 per cent alcohol and transported to the herbarium or taxonomy laboratory. A 30 per cent formaldehyde solution can also be used to preserve specimens before drying. The tag on fruits samples should contain the same data of the samples from the fertile branches, including the same number of collector.

## Collection Sites and some Species

**Figure 5.1**
A: Campo Rupestre in Goiás, Brazil; B: Cerrado in Goiás, Brazil after Bush Burning.

**Figure 5.2**
A: *Mandevilla ilustris* (Vell.) Woodson – Apocynaceae;
B: Fruits of *Cabralea sp.* – Meliaceae; C: *Lantana sp.* – Verbenaceae.

**Figure 5.3**
A: *Hyptis* sp. – Lamiaceae; B. *Lippia* sp. – Verbenaceae;
C: *Jacaranda ulei* – Bignoniaceae.

**Figure 5.4**
A: *Eriope* sp. – Lamiaceae; B; *Aspilia* sp. – Asteraceae;
C: *Palicourea rigida* – Rubiaceae.

**Figure 5.5**
A: *Hyptenia* sp. – Lamiaceae; B: *Caryocar brasiliense* - Caryicaraceae;
C: *Hypenia* sp. – Lamiaceae Rubiaceae.

## 5.3 Bench Works in the Herbarium

The activity involving the samples from field in the herbarium are focused towards the preservation of botanical materials for future references. The pressing and drying are two important steps in preserving plant specimens. Generally, pressing often prevents plant parts from curling or wrinkling during the drying process, and allows plant parts to be easily identified. The pressing process involves laying the plant specimens in folded sheets of newsprint and placing them in a pressing frame prior to their tightening with straps. Drying involves an adequate length of time and exposure to dry air, and maintenance of the specimens in the press. A low ambient humidity and good airflow could ensure rapid and thorough drying of plant material. As the specimens dry, it may be necessary to further tighten the straps on the press to minimize shrinkage and wrinkling. Rapid drying promotes the best retention of plant colour, but excessively high temperatures or long drying periods can result in blackened, discoloured, and brittle specimens. For instance, the newsprint can be changed to speed up the drying process (British Columbia Ministry of Forests, 1996). The samples are later transferred to the taxonomy laboratory for preparation of herbarium specimens. The specimens are mounted on folders of cardboard, 30cm x 40cm, by sewing or gluing. Mounting provides physical support that allows the specimen to be handled and stored with a minimum of damage. The mounting equipment includes acid-free mounting paper, cardboard sheets, glue, plastic squirt, wooden spacer blocks, weights, needle and linen thread. A label containing all information about the specimen, place of collection, date of collection,

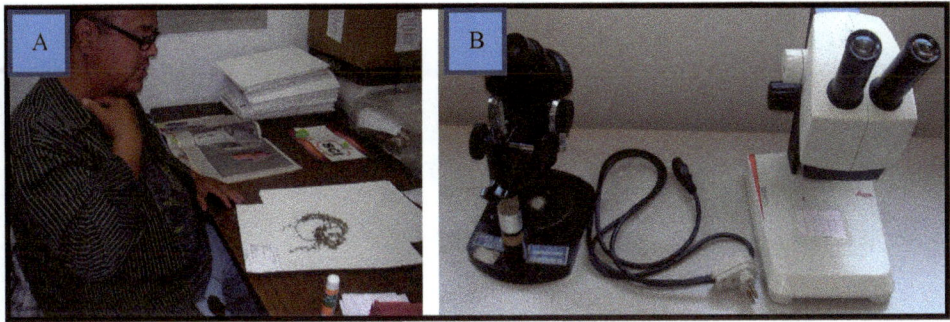

**Figure 5.6**
**A: Mounting of voucher specimen; B: Microscope - Stereoscopes.**

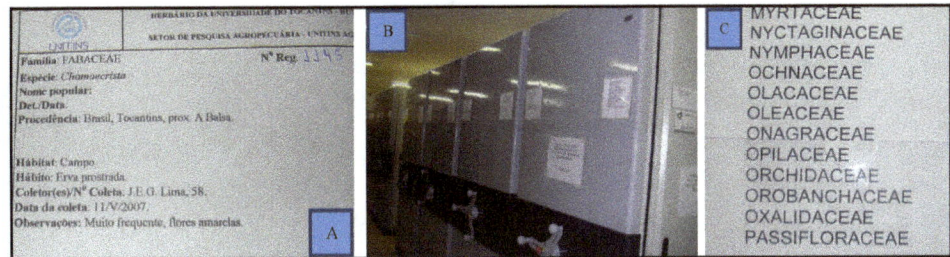

**Figure 5.7**
**A: Detailed label of voucher specimen; B: Dried specimen cabinets;**
**C: Plant families.**

collector number, the collection site [country, state, county, a geographical features (such as mountain, river etc)], geographical coordinates, the habitat of the plant (forest, savannah, swamp, path, rocky fields, etc), plant height, leaf colour, flower, ripe fruit and trichomes (as these structures or phenotypes could be lost over time) must be affixed at the bottom of the voucher specimen' s cardboard.

# Chapter 6

# Methods of extraction, isolation and Standardisation of Phytochemicals

*Christianah A. Elusiyan*

## ABSTRACT

*Metabolites obtained from various natural sources including plants (either whole or part), microbes, animals and even marine organisms remain sources of interesting biomolecules as candidate agents of therapeutic potentials. In order to achieve the optimal extraction and yield of active phytochemicals, correct processing, extraction and preparation of the materials must be followed using established procedures. Various chromatographic methods of analysis including newer hyphenated techniques have proven to be suitable for the separation and isolation of different classes of bioactive agents. Structural determinations are often achieved by spectroscopic studies and other qualitative and quantitative methods of analysis.*

*Lack of standardisation of medicinal product still remains a major criticism levelled against its usage especially in developing countries. Standardized protocols must be established and followed at each stage of preparation of any herbal remedy.*

## 6.1 Introduction

The subject of natural product chemistry or investigation of naturally occurring substances including plant materials is all encompassing. There are several reasons for studying natural products. The fact remains that they are the most common resource available locally as therapeutic agents. Many more indigenous traditional people in developing countries and even in the developed world rely on natural products including medicinal plants as food and medicine. It has been established that only about 10 per cent of plants available worldwide had been subjected to any investigation for biologically active compounds. So the wild still hold great promise of novel biomolecules waiting to be tapped as candidate drugs. Also there are other organisms in our fauna and flora that are not fully developed locally with respect to collection and taxonomy. An impressive number of modern drugs have been derived from natural sources and over the last century, a number of top selling drugs have been developed from natural products. However, there are still lots more to be explored and exploited for the benefits of mankind. Furthermore, they also serve a template plates for the syntheses of newer, less toxic or more potent derivatives.

Natural products are composed of products obtained from different sources including plants, microbes and animals including marine organisms. According to the World Health Organisation (WHO), some 3.4 billion people in the developing world rely on plant or plant based medicines for their various health needs including both infectious and non-infectious diseases. This represents about 88 per cent of the world's population, who rely solely on traditional form of medicine for their primary health care.

There are more than 400,000 known plant species around the world; these produce astonishingly diverse, complex, interesting and therapeutically array of chemical substances.

The aim of phytochemical studies is to isolate or purify compounds and identify them, whereas in analytical work the goal is to get information about the sample, determine the chemical profile of the metabolites.

Generally, plants are composed of naturally occurring substances known as metabolites. Metabolites are the intermediates products of metabolism. The term, metabolite, is usually restricted to small molecules (molecular weight < 1500 amu), produced by living organism, but which are not exactly essential for the survival and growth of the organism. Rather, they are biologically active chemical compounds with specific therapeutic properties. Interestingly, many of today's human or veterinary medicines or lead compounds are obtained directly or indirectly from natural sources. This variety of compounds can be divided into primary metabolites and secondary metabolites.

Primary metabolites refer to substances that are produced by plants and are essential for the short-term survival of the plant. They are involved in the proper growth, development and reproduction. These central key components are essential for maintenance of normal physiological processes occurring in organisms. Such substances include proteins, amino acids, fatty acids, carbohydrates and nucleotides.

Secondary metabolites are typically organic substances produced by plants through modification of primary metabolites. Secondary metabolites do not play a role in the growth and development and even reproduction of organisms and may not be essential or necessary for the primary biochemical activities and survival of plant and were once believed to be waste products. However, such substances are known to confer in many cases some kind of evolutionary advantage. For example, many secondary metabolites usually have important ecological function and therefore serve as chemical defence compounds against herbivores or infection, or they are involved in other types of interactions with different organisms (such as attraction of pollinators or allelopathy). Many secondary plants metabolites are drug substances or serves as template for the development of newer medicines which are therapeutically valuable entities. For instance, Erythromycin, derived from

*Saccharopolyspora erythraea,* is a commonly used antibiotic with a wide spectrum of antimicrobial activity. Some plants contain toxic secondary metabolites and caution must be exercised when such are being investigated. For instance all the parts of the English yew plant, *Taxus baccata,* contain poisonous substances.

Plant's secondary metabolites are distributed in the whole plant and various parts or organs of plant including the leaves, roots, stems, fruits, seeds, rhizomes and even exudates. Other natural sources that generate metabolites are microbes and animals including marine organisms. Plants have provided humans with medicines and this practice has been in existence since time immemorial. Many drugs currently in use today are plant derived, such as opioids (including morphine and codeine), anti-cancer agents such as paclitaxel (Taxol®) and vincristine, and galanthamine, which is used for the treatment of Alzheimer's disease. Similarly, microorganisms have yielded many important antibiotics, for instance, Bacitracin, derived from organisms classified under *Bacillus subtilis* is an antibiotic commonly used in clinical practice as a topical agent. Other species including fish and other marine organisms are the source of fatty acids with important health benefits.

The metabolites display great and rich chemical diversity and are represented by many different classes of naturally occurring compounds. Major groups of phytochemicals of therapeutic interest are alkaloids, terpenes, including volatile oils (which are composed primarily of monoterpenes and sesquiterpenes) and fixed oils; phenolic compounds including flavonoids and tannis; and a range of glycosides with different aglycone (non-sugar) moieties.

## 6.2 Some Major Groups of Phytochemicals of Therapeutic Interest

### Alkaloids

Alkaloids are naturally occurring secondary metabolites. They are organic bases at least a nitrogen atom within a ring system. Alkaloids often have profound physiological action on man and animals and are widely used in medicine. They are biosynthetically derived from amino acids and are found at 10-15 per cent concentration in almost all plants. Some alkaloids occur in plants in the free state while majority are present as soluble salts. Alkaloidal bases are generally soluble in lipophilic organic solvents such as dichloromethane but are insoluble in water but the salt forms are soluble in aqueous or polar organic solvents but insoluble in lipophilic solvents. Some alkaloids can be extremely toxic. **Morphine** was the first alkaloid to be isolated from *Papaver somniferum',* or the opium poppy. It is used as a pain reliever in patients with severe pain levels and cough suppressant. Another example of an alkaloid is **cocaine**. It can be highly dangerous and addictive. However, it has also been used as an anaesthetic. Cocaine derivatives are very dangerous when habitually used and can be deadly. The most known alkaloid is **caffeine.** It is found in cocoa, coffee and tea and it is used to stay alert and also possesses some protective properties on the plants from which it is obtained. Seedlings of the coffee plant have a high concentration of caffeine. The high concentration is toxic and protects the seedlings from insects that want to prey on it.

Most alkaloids have a bitter taste. It is believed that plants evolved the ability to produce these bitter substances, many of which are poisonous in order to protect themselves from predators *e.g.* animals (grazing), infections, harsh weather/climatic conditions, etc; for example, aporphine alkaloid liriodenine produced by the tulip tree protects it from parasitic mushrooms.

## Glycosides

A glycoside is a non-reducing organic compound which on hydrolysis yields an aglycone part (sometimes called the genin), nonsugar component and one or more reducing sugars. Generally, the physical and chemical properties of a glycoside is determined by the chemical nature of the aglycone, *e.g.* terpenoid, flavonoid, etc. It also depends on the type of linkage between the aglycone and the sugar moiety, *e.g.* C-O-C, C-S-C, C-N-C, C-C-C linkage. It is also determined by the number and type of sugar residues in the molecule. The more the sugar residues present in a glycoside the greater its water solubility. Glycosides are soluble in water and alcohol but insoluble in organic solvents. The aglycone, however, is soluble in organic solvents while the sugar portion is insoluble. As a result of this, the isolation of the components of the hydrolysed glycoside requires solvents different from those used for the isolation of the glycoside. Glycosides are best isolated using reverse phase chromatography eluting with aqueous solvent systems.

## Terpenoids

**Terpenoids** are made of isoprene units and are found in all plants. They are the largest group of secondary metabolites and are very volatile, which means they evaporate easily and are thermally unstable.

**Essential oil** or volatile oil gives plants their definite fragrance. In some plants the scent is used to deter herbivores and protect the plant from dangerous pathogens. A volatile oil is a naturally occurring mixture of volatile constituents which usually consists of mixtures of terpenes, which are unsaturated hydrocarbon with general formula $(C_5H_8)n$. They can further be classified as alcohols, esters, aldehydes, ketones and phenols, etc. Volatile oils are soluble in organic solvent and are partially soluble in alcohol but insoluble in water. Fixed oils are naturally occurring non-volatile oily substances consisting of glycerides (glyceryl esters of long-chain fatty acids) with small quantities of steroidal substances. They are insoluble in water and alcohol but soluble in organic solvents.

We use essential oils for aromatherapy and medicine. In aromatherapy, essential oils are thought to improve the mood and mental functioning. Most of the time, essential oils are dangerous if consumed so they are usually applied topically or inhaled. They can be used for skin infections, respiratory ailments, as antiseptics and insecticidal agents.

## Phenolic Compounds

Phenolic group of compounds is a large group of structurally diverse naturally occurring compounds that possesses at least a phenolic moiety in their structures. The group encompasses various structural types including phenyl propanoids, coumarins, flavonoid and isoflavonoids, lignans and tannins, etc.The phenols consist of a **hydroxyl group** (–OH) attached to an aromatic ring. They are widespread in plants families. They have become very popular due to their health benefits as antioxidants.

Tannis are complex phenolic substances, soluble in water and alcohol, having an astringent and bitter taste.

Flavonoids are the largest group of naturally occurring plant phenols, occurring both in the free state and as glycosides and they are frequently coloured substances. The glycosides are soluble in water and alcohol but insoluble in organic solvents; the genins are only sparingly soluble in water but are soluble in ether and other non-polar solvents and to a certain extent in alcohol. Flavonoids can be further divided into three groups: anthocyanins, flavones and flavonols.

**Anthocyanins** range in colour from red to blue and purple. The colour depends on the pH of the environment. Anthocyanins are most commonly found in grapes, berries and have a wide range of health benefits. Anthocyanins are believed to protect against heart disease, diabetes and even cancer when they are consumed. They are also appearing in skincare products to slow down the aging process.

**Table 6.1: Some Secondary Metabolites of Medicinal Importance**

| Secondary Metabolites | Class of 2 Metabolites | Biological Source | Family | Uses |
|---|---|---|---|---|
| Penicillin | Antibiotics | *Penicillium griseofulvum* | Peniciliinacea | Antibiotic |
| Quinine, quinidine | Alkaloid | Cinchona bark (*Cinchona succirubra*) | Rubiaceae | Anti-protozoal, Bitter tonic, anti-arrythmic |
| Ergometrine, ergotamine | Alkaloid | *Claviceps spp* | Clavicipitaceae | Uterine stimulant |
| Vincristine | Alkaloid | *Vinca rosea* | Apocyanacea | Antileukemic |
| Morphine | Alkaloid | *Papaver somniferum* | Papaveraceae | Narcotic analgesic |
| Artemisinin | Sesquiterpene | *Artemisia annua* | Asteraceae | Antimalarial |
| Taxol | Alkaloid | *Taxus brevifolia* | Taxaceae | Anticancer drug |

# 6.3 Material Collection, Processing and Extraction Procedures

## Collection

Phytochemical study of natural products often begins with collection of material. Collection at the right time, season and location must be ensured. The collected material either plant or animal must be properly identified by qualified personnel because examples of misidentification, adulteration, and contamination of plant based products have been widely recorded. Possibly, voucher specimen must be kept stored for future reference in a local herbarium.

Intact or whole plant parts should be collected and not infected materials. Collect the flowers and the herbs when they are in full bloom, while leaves, grasses, aerial parts should be collected before they begin to turn brown. Barks and gum resins should be obtained from mature trees.

Caution should be exercised when collecting barks and roots of plants to prevent injury and eventual death of the plant. Do not collect round the entire trun of the tree. Surface roots and not deep root should be collected. Time of collection should be well considered. Seasonal variations in composition of plant constituents as well as biological or therapeutic effects have been widely reported. In the study of two Brazilian species; *Guapira graciliflora* and *Pseudobombax marginatum*, the content of total polyphenols was found to be higher in winter for *P. marginatum* and in summer for *G. graciliflora* while the opposite was found for the flavonoids composition of the plants (Chaves, *et al.*, 2013).

## Processing

After collection, the material must be processed appropriately. To prevent enzymatic actions and deterioration, the collected materials can be subjected to drying especially if processing and analysis is not immediate. Open air drying at average room temperature of 20-40°C is highly recommended over a couple of days for leaf materials. Aromatic plants, as much as possible, should be processed fresh to prevent loss of volatile constituents especially the lower monoterpenes. Rapid drying using hot air ovens and drying rooms fitted with exhaust fans can also be used for drying flowers and herbs at 20-40°C, while root and barks can be dried at 35-65°C. Drying time may increase or decrease depending on the atmospheric temperature, humidity, moisture content of the material.

Other forms of drying includes solar dryer in tropical area, however, direct sun drying and other high temperature drying facilities must be avoided. Materials must be dried until a constant weight is attained.

## Extraction

Extraction refers to the process of separation of metabolites portions of either plant or animal tissues from other inactive or inert components by using established extraction procedures. The crude extractives so obtained from these natural sources are often complex mixtures of substances of varying polarity and relatively impure liquids, semisolids or powders. Depending on the mode of extraction, the materials often include different classes of preparations known as decoctions, infusions, fluid extracts, essential oils, tinctures, extracts (semisolid in nature) and powdered extracts.

The purposes of standardized extraction procedures for crude drugs are to attain the therapeutically desired portion and to eliminate the inert material by treatment with a selective solvent known as *menstrum*. The specific method to adopt for extraction is often influenced by many factors. The purpose of extraction, the intended use of the extracted material, nature of the extractive, are some of the factors that could be considered in order to achieve the uttermost result and yield. The choice of solvent or mode of extraction is sometimes determined by the use in ethnomedicine; *i.e.* mimicking the traditional processes as much as possible, "use it the way it is being used" in order to achiev the desired therapeutic effect.

The extracted material (standardised extract in this case) may then be utilised directly as a medicinal agent in the form of tinctures and fluid extracts. Therefore, standardization of entire procedures starting with plant collection, to processing and extraction contributes significantly to the final quality of the herbal medicinal drug.

The extract so obtained may also be further processed and incorporated in any dosage form such as tablets, capsules, creams or ointments, or it may be fractionated and further separated in order to isolate the individual chemical substances using various chromatographic techniques. Identities of these isolated chemical entities such as artemisinin, hyoscine, digoxin and vincristine, which are modern drugs, are established by combination of various spectroscopic methods of analysis.

Some extraction methods are as follows:

1. Maceration
2. Percolation
3. Infusion
4. Decoction
5. Digestion
6. Hydro-distillation : aromatic plants meant for extraction of its essential oils must be processed fresh to prevent loss of the vital monoterpene components during drying.
7. Solvent extraction
8. Hot continuous extraction – Soxhlet extraction
9. Supercritical Fluid Extraction

The aim of any of the extraction process is to enable the released of active metabolites from the plant or animal tissue. So therefore, the method of choice must be appropriate, suitable and must give the most yield of the extractive. For standardisation and for the purpose of quality control, some parameters to be considered for choosing the right extraction process are below.

## Important Factors to be Considered for Selecting an Appropriate Extraction Method

1. Authentication and identification of plant material should be done by qualified personnel before performing extraction. Get rid of any foreign matter and wash in running water if necessary.
2. Collect the right plant part and, for quality control purposes, record the age of plant and the time, season and place of collection.
3. Conditions used for drying the plant material largely depend on the nature of its chemical constituents. Hot or cold blowing air flow for drying is generally preferred. If a crude drug with high moisture content is to be used for extraction, suitable weight corrections should be incorporated. The material should be dried to a constant weight and the moisture content determined.
4. Grinding methods should be specified and techniques that generate heat should be avoided as much as possible. Powdered plant material should be passed through suitable sieves to get the required particles of uniform size.
5. Nature of *Constituents:* a) If the active biomolecules are non-polar constituents in nature, a non-polar solvent may be used for extraction. For example, lupeol is the active constituent of *Crataeva nurvala* and, for its extraction, hexane is generally used. Likewise, for plants like *Bacopa*

*monnieri* and *Centella asiatica*, the active constituents are glycosides and hence a polar solvent like aqueous methanol may be used. b) If the constituents are thermolabile, extraction methods like cold maceration, percolation and hyrdrodistillation are preferred. For thermostable constituents, Soxhlet extraction (if non aqueous solvents are used) and decoction (if water is the menstruum) are useful. c) Suitable precautions should be taken when dealing with constituents that degrade while being kept in organic solvents, *e.g.* flavonoids and phenyl propanoids. d) In case of hot extraction, higher than required temperature should be avoided. Some glycosides are likely to break upon continuous exposure to higher temperature, moisture and light.

6. Standardization of time of extraction is important. inadequate time means incomplete extraction. Longer than necessary time for extraction will lead to extraction of unwanted material. For example, if tea is boiled for too long, tannins are extracted which impart astringency to the final preparation. f) The number of extractions required for complete extraction is as important as the duration of each extraction. The quality and quantity of water or menstruum used should be specified and controlled. Likewise the quantity of the extracted material as well as the recovered extract and yield obtained should be noted. The design and material of fabrication of the extractor are also to be taken into consideration.

7. Concentration and drying procedures should ensure the safety and stability of the active constituents. Drying under reduced pressure (*e.g.* using a Rotavapor at 40°C) is widely used. Lyophilization, although expensive, is increasingly employed.

8. Analytical parameters of the final extract, such as TLC and HPLC fingerprints, should be documented to monitor the quality of different batches of the extracts.

## 6.4 Separation Techniques

Natural products are composed of complex mixtures of different classes of phytochemicals and to separate them into individual component is indeed a huge task. Various methods of purification (or "clean-up") may be used. For instance, the initial fractionation of the extract may be performed by partitioning between two immiscible solvents with the desired phytochemicals dissolving preferentially in one and not the other. Other reagents may be added that will selectively precipitate out either the desired compounds or the unwanted constituents leaving the compound(s) of interest in solution. The quantitative and qualitative analyse of natural products are mostly performed by chromatography. Chromatographic techniques are probably the most versatile and effective methods for analytical separation of chemical mixtures of plant extracts, oils and products.

All chromatographic methods are based on the same principle. The principle is such that each component in a sample mixture ordinarily interacts with its environment in a different way from all other species present in the mixture when subjected to the same condition.

Separation of the various components of the mixture is made possible by the distribution between two phases known as mobile (moving) and stationary (static) phase. When the mobile phase is a liquid, the technique is a called liquid chromatography (LC) and when it is a gas it is called gas chromatography (GC).

## Thin Layer Chromatography

Thin layer chromatography (TLC) is a very simple, yet useful analytical separation technique. The aims of TLC analysis include to determine the components of a mixture qualitatively; to compare and contrast the differences between plants species and; to compare components of an extract with pure reference compounds or authentic and standardised extract.

The sample to be analysed is made into solution and is then applied on the TLC plate and developed in appropriate solvent system known as the mobile phase (which could either be a single solvent or mixtures of solvents). The TLC plate will be examined to know if the components of the mixture have been resolved by using appropriate spray reagents to detect the separated spots.

There are many available TLC visualisation reagents specific and suitable for the detection of compounds or class of molecules. Before spraying, plates should be well dried of residual solvents. Plates should be visualised under normal white light and also at short and long wavelength (254nm and 366nm) UV light prior to spraying. Spraying of any reagent onto plates should be performed in well ventilated hood while wearing protective safety glasses. After spaying, some reagents require further treatment such as heating at a particular temperature to reveal the spots.

TLC visualisation reagents can be found in the EMD Chemicals catalogue or on the website (www.emdchemicals.com).

The retardation factor (Rf) values of each spot should be determined in a specific solvent system. The colour of the separated spots/bands should be noted. Rf values are defined as the distance traveled by substance from origin (or point of application) divided by distance moved by the solvent. As the substance travels with the solvent, the solvent will always have a greater or equal value to the distance travelled by substance. This means that the Rf value is usually equal to unity or less.

Another form of TLC is known as preparative TLC or simply Prep-TLC. It is a quantitative analytical technique used for isolation of compounds of interest. Instead of just examining the TLC plate for visible spots, the band corresponding to a particular spot can be scraped off and the substance extracted by eluting with the right solvent in which the substance is soluble in. The same mobile phase used to achieve separation on analytical TLC is the most appropriate for Prep-TLC but the plate should be 20cm x 20cm of adsorbent thickness layer of between 0.5 mm and 1 mm.

## Liquid Chromatography

In liquid chromatography, separation is achieved by passing a dilute solution of the sample or by applying the sample already pre-adsorbed on silica gel to the column packed with solid adsorbent particles which may or may not be coated with another liquid. Desired separation is often achieved by applying proper operating conditions, right choice of eluting solvents, and packings, some components of the sample mixture will move through the column while others movement of others will be retarded. The faster ones will elute first and be collected, while those that move more slowly will elute much later.

The 'conventional' liquid chromatography is performed by the use of vertical glass column filled with stationary phase and the mobile phase (either single solvent or mixtures of solvents) is allowed to travel down the column by gravity flow. Eluate or fraction collected is further examined by thin layer chromatography or by other techniques. In order to achieve good column fractionation, the sample mixture must have been well resolved or separated first on TLC. This often times requires a trial and error check in order to determine the solvent(s) with best resolving power prior to column separation. Isocratic (single solvent) or gradient (mixtures of solvents of increasing polarity) elution can be employed to elute the column. The fractions are collected and evaluated using TLC. Purity checks of the collected fractions and isolated compounds in at least 2 -3 different solvent systems is important in order to ascertain their purity status. Monitoring of isolation of bioactive constituents is made possible by using suitable bioassays test procedures and this is referred to as bioassay-guided isolation. General screening method measuring different types of biological/pharmacological activity could be used. The biological test screening method must be simple and rapid to accommodate large number of semi-purified

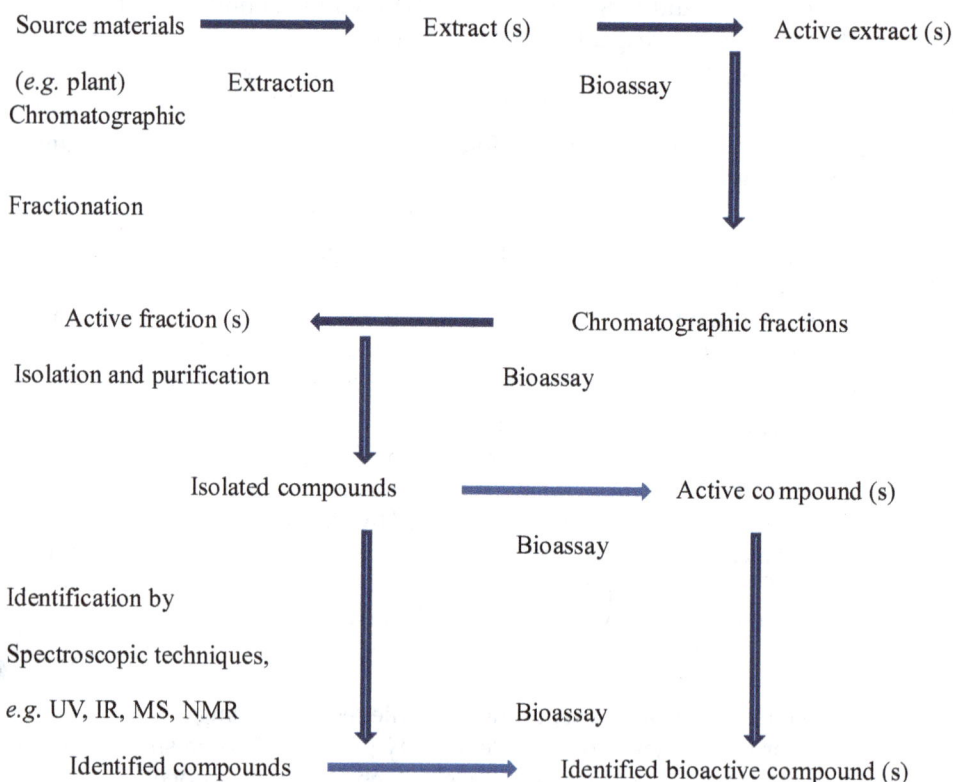

**Figure 6.1: Scheme of Bioassay-Guided Isolation Processes.**

or purified fractions to be screened. Solubility is a critical problem that must be overcome when performing biological tests; therefore appropriate solubilizing or emulsifying agents must be used to carry and deliver the active substances to the target. Examples of simple bioassays are Hole-in plate/MIC determination for microorganisms, various bioautographic techniques, brine shrimp cytotoxicity test and others.

Though LC is less sophisticated, it's usually a slow separation technique. In spite of this demerit, it has enabled many remarkable separations to be achieved. Other variants of the LC including high pressure or high performance liquid chromatography (HPLC) have been developed to improve its speed and versatility. In addition, separations of complex, in volatile and very polar substances have been achieved. By choosing the right column packing material, normal phase, reverse phase, ion-exchange or gel filtration separations may be effected by using the HPLC technique.

## Gas Chromatography

In gas chromatography (GC), the mobile phase is a gas, and the stationary phase is a liquid coated on an inert solid support. It is pre-packed into the column through which the gas is allowed to flow through. Separations of the component of mixture are achieved by partitioning of the components between the gaseous mobile phase and the liquid stationary phase at different times known as the *retention time*. Components eluting from the column are detected and the detected signal transformed into a peak. Quantitative analysis of the components can be achieved by measurement of the areas under each peak by using internal standard.

For effective analysis, chemical substances for GC must be volatile. GC is the most commonly used method to selectively detect, identify, and quantify the volatile hydrocarbon components of essential oil obtained from aromatic plants. Generally, the type of plant material determines which method to be used to obtain the essential oil. The extraction method used determines the quality of the oil that is produced; since a wrong or wrongly executed extraction, can damage the oil, and alter the chemical nature of the essential oil. Oils can be extracted by different methods including hydrodistillation using clavenger-type apparatus, steam and water distillation, expression, maceration, organic solvents and supercritical $CO_2$ extraction. Steam distillation is one of the oldest and most acceptable methods of extracting essential oils. Although hypercritical $CO_2$ is a great way to extract most oils because it produces oils of superior quality, but the cost involved in makes this method highly unaffordable by most people. To achieve effective extraction and to prevent loss of the monoterpene components of the essential oil, it is advisable to use fresh plant tissue materials. Hydrodistilled plant materials should initially be extracted for 4 hours using Clavenger apparatus and recover the volatile oils. The plant material can further be hydrodistilled for another two hours to recover more components. Collected oils should be immediately died over anhydrous sodium sulphate and stored refrigerated.

## 6.5 Chromatographic Fingerprinting and Marker Compound Analysis

Chromatographic fingerprint is an innovative approach for the identification and authentication chemical biomarkers of natural products and Herbal Medicine (HM). The chromatographic pattern of the components of the extract is established by instrumental fingerprinting. Chromatographic fingerprinting profile for single herb, herbal ingredients and finished products are established using methods such as High-Performance version of Thin Layer Chromatography (HPTLC), High-Performance Liquid Chromatography (HPLC), Gas Chromatography (GC), Liquid Chromatography Mass Spectrometry (LC-MS) and Gas Chromatography (GC-MS).

Chromatographic fingerprinting techniques are playing very important role in the standardisation and quality control of herbal products. They are being used for identifying and confirming the presence of different herbs in poly herbal formulations, quantification of marker compounds as well as the determination of quality and purity of the formulated product and detection of adulteration.

With the help of chromatographic fingerprints so obtained, the authentication and identification of herbal medicines can be accurately known. Herbal formulation and any plant extract are generally complex mixture containing several unknown components and many of them are usually present in minute quantity. Moreover, there usually exists variability within the same herbal materials collected at different season or different geographic locations. It is therefore very important to obtain reliable chromatographic fingerprints that represent pharmacologically active and chemically characteristic components of the product or ingredient. To achieve this type of phytochemical evaluation of herbal drugs, TLC has been employed extensively for the certain reasons including its ability to enable rapid analysis of herbal extracts without much sample preparation and extensive sample clean-up. Secondly, TLC provides qualitative and semi quantitative information of the resolved compounds and it enables the quantification of chemical constituents. Fingerprinting using HPLC and GLC is also carried out in specific cases. TLC fingerprint profile of the sample can be recorded using a HPTLC scanner. Data obtained would include the chromatogram, the retardation factor (Rf) values (in a specified solvent system), the colour of the separated bands, their absorption spectra (at maximum wavelength) and shoulder inflection(s) of all the resolved bands amongst other profiles. Information on the chemical profile so obtained helps to validate the authenticity and quality of the herbal drug and the ingredients and to eliminate unwanted adulterants.

HPLC fingerprinting provides information on the retention time of individual peaks present in the mixture, records the chromatograms using different mobile phases. In a similar way, GLC generates fingerprint profiles of volatile oils and fixed oils of herbal drugs. Furthermore, the recent approaches and developments of hyphenated chromatography and spectrometry techniques such as Gas Chromatography–Mass Spectroscopy (GC–MS), Liquid Chromatography–Mass Spectroscopy (LC–MS) High-Performance Liquid Chromatography–Diode Array Detection (HPLC–DAD), Capillary Electrophoresis- Diode Array Detection

(CE-DAD), and High- Performance Liquid Chromatography–Nuclear Magnetic Resonance Spectroscopy (HPLC– NMR) could provide the additional spectral information, which will be very helpful for the qualitative as well as the quantitative analysis of the herbal material and even for the on-line structural elucidation of the chemical markers contained therein. (Liang, *et al* 2004, Patil and Shettigar 2010 and Nikam *et al*, 2012)

## 6.6 Dereplication or Chemical Screening

Dereplication is the process by which recurrence or re-isolation of same or similar compounds from various extracts can be eliminated in the shortest time

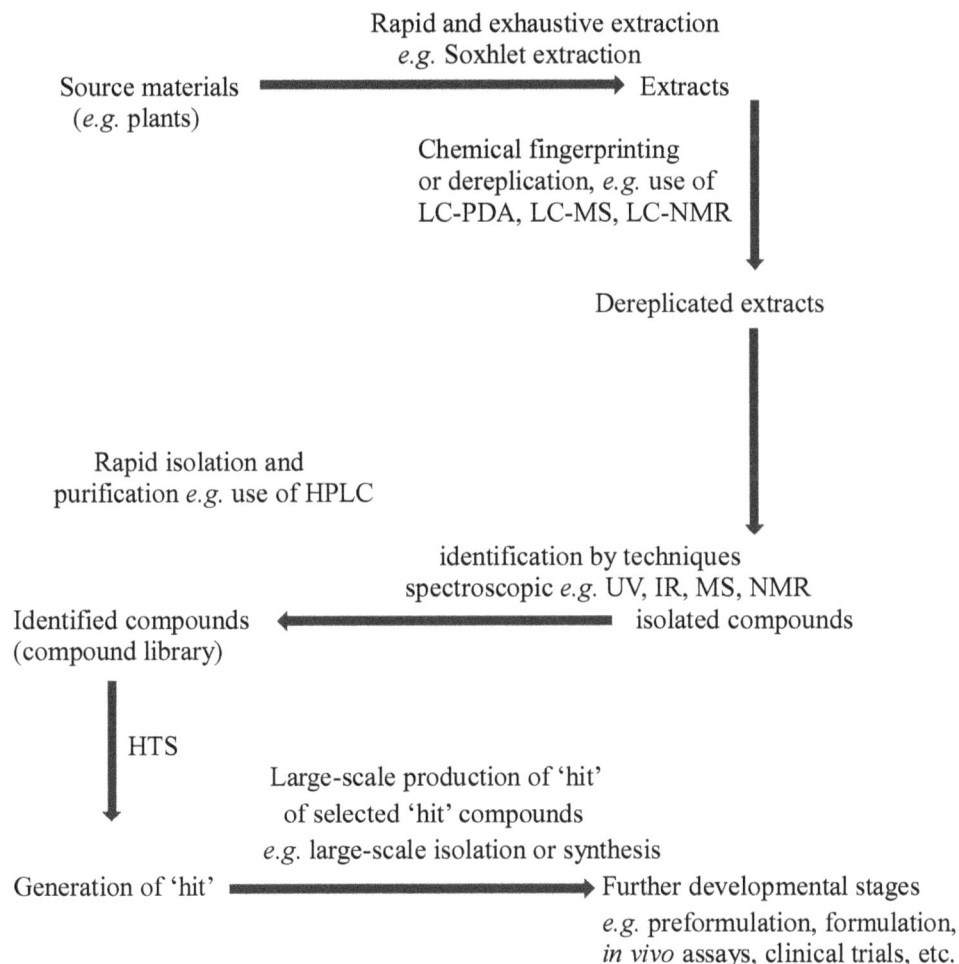

Rapid and exhaustive extraction
*e.g.* Soxhlet extraction

Source materials   ━━━━━━━━━━━━▶ Extracts
(*e.g.* plants)

Chemical fingerprinting
or dereplication, *e.g.* use of
LC-PDA, LC-MS, LC-NMR

Dereplicated extracts

Rapid isolation and
purification *e.g.* use of HPLC

identification by techniques
spectroscopic *e.g.* UV, IR, MS, NMR

Identified compounds  ◀━━━━━━━━━ isolated compounds
(compound library)

HTS

Large-scale production of 'hit'
of selected 'hit' compounds
*e.g.* large-scale isolation or synthesis

Generation of 'hit'  ━━━━━━━━━━▶ Further developmental stages
                    *e.g.* preformulation, formulation,
                    *in vivo* assays, clinical trials, etc.

**Figure 6.2: Scheme of Modern High Throughput Screening (HTS)
Screening Processes.**

possible. The approaches involve High Throughput Screening (HTS) which helps to avoid time wasting involved in isolation of common or known constituents.

Using several hundreds of substances can be screened using several assays within a short time, and with very little quantity of compounds. A number of hyphenated techniques are used for dereplication, *e.g.* LC-PDA (liquid chromatography–photo-diode-array detector), LC-MS (liquid chromatography–mass detector) and LC-NMR (liquid chromatography– nuclear magnetic resonance spectroscopy). It also involves exhaustive literature search (with proper acknowledgement of previous authors). The modern HTS programmes incorporate a 'high quality' and 'chemically diverse' natural product library (a collection of dereplicated natural products). Natural product libraries can also be of crude extracts, chromatographic fractions or semi-purified compounds. However, the best result can be obtained from a fully identified pure natural product library as it provides opportunity to handle the 'lead' molecule rapidly for further development, *e.g.* toxicity tests, total or partial synthesis, formulation studies, *in vivo* assays and clinical trials.

## Chapter 7

# Preclinical Techniques for the Evaluation of Psychoactive Plants

*James O. Fajemiroye and Elson A. Costa*

## ABSTRACT

*Preclinical evaluation of psychoactive plants has revealed a variety of promising medicines that may provide benefit in the treatment of wide varieties of mental related disorders. This chapter reviews some common animal models and the rationale behind their uses. An open-ended search of PubMed, Scopus among other electronic databases was conducted using relevant search criteria and keywords. Numerous species including Sonchus oleraceus, Stachys lavandulifolia, Cecropia glazioui, Eschscholzia californica, Rubus brasiliensis, Apocynum venetum, Securidaca longepedunculata, Achillea millefolium, Coriandrum sativum, Eurycoma longifolia, Turnera diffusa, Euphorbia hirta, Crocus sativus, Aloysia polystachya and Casimiroa edulis have been subjected to preclinical tests. Although clinical trials remain the ultimate step in new drug discovery, preclinical screening are essentially vital towards the detection of psychoactive plants with therapeutic potential.*

## 7.1 Introduction

Crude extract from psychoactive plants or phytoconstituents are often subjected to preliminary screenings otherwise known as general pharmacological tests, hippocratic or Irwin test to determine if plant extracts or its isolate possess biological activity or not. The group number of animals (n) in this study is generally small. Effects of extract/compounds on CNS or autonomic nervous system can be observed and scored using different parameters like stimulant (hyperreactivity, irritability, aggressiveness, tremors, seizures, piloerection, intense movement of whiskers, etc) and depressant (ptosis, sedation, loss of ear reflex, anesthesia, ataxia, loss of righting reflex, catatonia, analgesia, touch response, loss of corneal reflex, hypnosis). Behavioural changes in ambulation, grooming, rearing, vocalization, writhing, paralysis of posteriror limb, stereotyping or some manifestations like salivation, diarrhea, muscle tone, paw strength, constipation, defecation, tears, urination, cyanosis, forced breathing through autonomic nervous system. In some cases, a toxic extract/compounds can lead to death of animals. The effects can be recorded on a scale like (0) no effect, (1) presence of effect while the intensity of the effects can be symbolized with sign like (-) decreased effect, (—) lower effect, (+) increased effect, (++) higher effect, etc.This study can also be used to investigate appropriate route of administration (p.o, iv, sc, ip,.) and estimate doses of botanicals that do not constitute risk of toxicity (Malone, 1983; Cunha *et al.*, 2009). A typical general pharmacological test was described in Irwin test (Irwin, 1962 and 1968).

## 7.2 General Pharmacological Tests - Irwin Test

This test is commonly used to evaluate the effects of a new substance on behavioural and physiological function. The results of the Irwin test are used to determine potential toxicity and to select doses for specific pharmacological activity. The Irwin test can also be used in a safety approach for detecting untoward effects of a new compound on general behaviour and for evaluating its acute neurotoxicity. In particular, data obtained in the Irwin Test can help to determine the dose range to be tested in other safety tests. Furthermore, the Irwin Test can furnish a first but pertinent orientation towards a specific therapeutic indication, a specific mechanism of action or a specific physiological function (Roux *et al.*, 2005). In order to conduct the Irwin test, mice are divided into different treatment groups (extract/compounds/ reference drugs/vehicle), each group consisting of 3-5 mice. The animals are treated ip, sc, iv, etc. and observed at different interval for minutes, hours, days, weeks or months to determine appropriate route of administration.

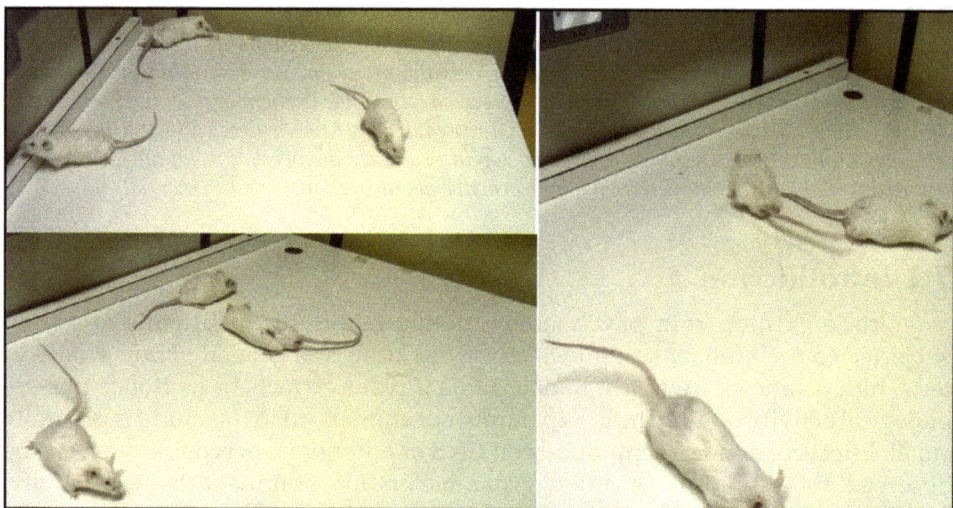

**Figure 7.1: Free Ambulation of Male *Swiss Albino* Mice (weight 25 g, 4 weeks) for Periodical Observation (5, 20, 30, 60, 4x60, 24x60, 24x60x7 minutes) of General Behavioural Changes in Irwin Test.**

## 7.3 *In vivo* Evaluation of CNS Depressant or Stimulant Effect

Psychoactive extracts or compounds could sometimes elicit general effects like CNS depression (depression of CNS does not indicate pathological depression) or stimulations that influence animal behaviour in a specific model. Groups of animals can be pretreated with barbiturate prior to the treatment with novel extracts or compound, reference drugs or vehicle that was used to dissolve drugs. Generally, the choice of route of administration determines the interval between pretreatment and treatment. It is advisable that the experimenter should avoid using the same route of administration for the pretreatment with barbiturate and treatment with novel drugs. An unknown compound could be assumed to possess CNS depressant effect if its potentiate barbiturate induced hypno-sedative effect as compare to the control group of animals. A reduction in the effect of barbiturate could suggest CNS stimulant effect of the novel extract or compounds. The time taken for the loss of righting reflex and the duration of sleep (time lapse between loss of righting reflex and voluntary recovery time). Alternatively, ethyl ether can be used to induce sleep (Carlini and Burgos, 1979).

### Barbiturate Induced Hypno-Sedative Effect

Mice are treated intraperitoneal with a known dose of sodium pentobarbital prior (at an interval of 1-h) to oral administration of vehicle (control group), extract/ phytoconstituents or diazepam (reference drug). The sleep latency (time taken for the loss of righting reflex) and duration (time lapse between loss of righting reflex and voluntary recovery time) are recorded and subsequently analysed. Potentiation

Figure 7.2: Sodium Pentobarbital (50mg/kg) Induced Hypnosedative Effect in Male *Swiss Abino* Mice (weight 25 - 30 g, 4 weeks).

(decrease in sleep latency and increase in sleep duration) or attenuation (increase in sleep latency and decrease in sleep duration) of hypnotic effect of sodium pentobarbital can be detected through the comparison of control group with the extract, phytoconstituents, diazepam treated group (Fajemiroye *et al.*, 2012; Carlini and Burgos, 1979).

## Diethyl Ether Induced Sleep

The mice are pretreated with vehicle, extract, phytoconstituents, reference drugs prior to their placement (individually) on a platform in a transparent glass chamber (30 cm x 20 cm diameter) saturated with ethyl ether until they lose righting reflex (sleep onset). The latency to loss of righting reflex and the total duration of sleep are reported. This method has an advantage over barbiturate-induced sleep, as it is not influenced by metabolic activities of hepatic enzymes that could constitute pharmacokinetic-induced false positive results.

## 7.4 Evaluation of Motor or Locomotor Activity

The evaluation of motor activity is important following treatment with organic extracts and phytoconstituents. Intoxication, sedation and incoordination motor can be observed phenotypically through animal model. However, alterations in motor/locomotor activity are not always mediated centrally. Wire hanging, rota rod and open field tests can be used to assess alteration in locomotion activities.

## Wire Hanging Test

The wire hanging test can be used to evaluate pharmacological effect of drugs (extracts and phytoconstituents) on motor function (motor impairment or coordination) of experimental animals. Animals are randomly picked and subjected to administration of vehicle, extracts, phytoconstituents or reference drugs. The

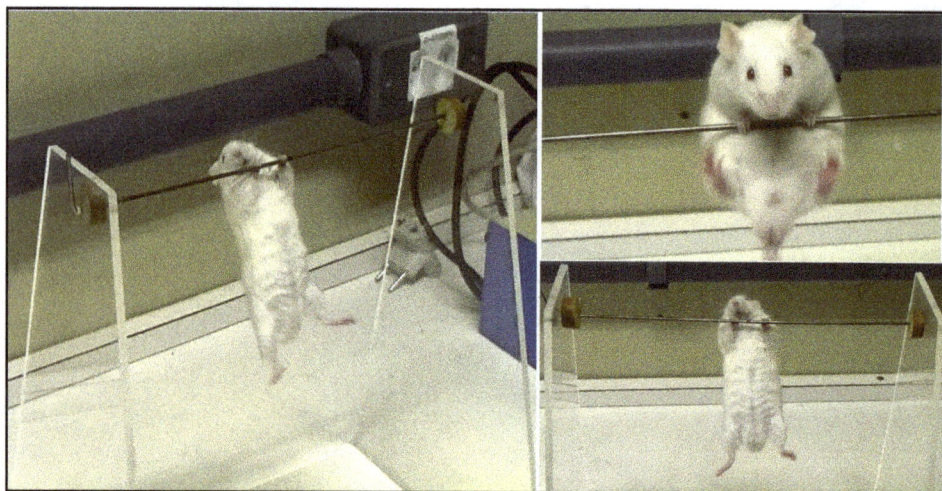

**Figure 7.3: Male *Swiss Abino* Mice (weight 25 g, 4 weeks) during Wire Hanging Session for 1 Minute (Apparatus specification/condition - diameter and length of wire...).**

test starts with the animal hanging from an elevated wire by their forepaws at a predetermined height that prevent the animal from climbing down. This test measures forelimb strength and coordination. The animal are placed at the centre of the wire. The time that elapsed until the animal fell can be recorded three times within the predetermined cutoff time. The number of falls and latency to the falls are important parameters in this model (Fajemiroye *et al.*, 2014). An increase in fall and decrease in latency to the first fall could suggest motor impairment.

## Rota Rod Test

The rota rod test is a well-established model originally developed by Dunham and Miya (1957) for the evaluation of motor coordination of animals. The latency to fall from a rotating rod is scored manually by experienced researcher or automatically. In this model, a constant revolution per minute (rpm) of rota rod that allows animals with poor coordination to fall off is maintained throughout the experiment. In addition to the motor coordination that can be observed at the beginning of the animal's exposure, the longer stay could be a measure of endurance. Also, cognitive aspects can be measured in this model. Animal could learn to drop off as soon as it is placed or kept on the rotating rod through implicit memory. This kind of memory can be assessed through repeated training of the animals for a period of time prior to testing the animals to see if the latency to fall when re-introduced is significantly lower than it was on the last trial. Hence, wide range of behaviours could be associated to non-parametric nature of the experimental data in this model. Motor coordination can be tested by comparing the latency to the first fall between experimental and control groups.

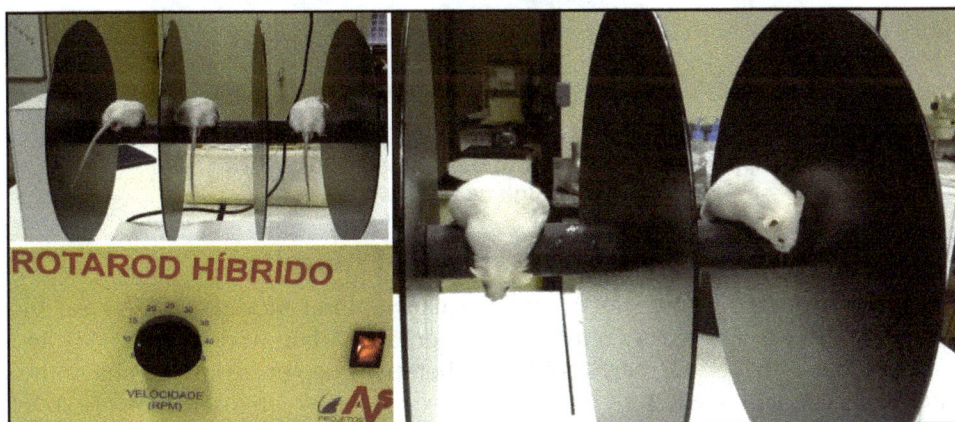

**Figure 7.4: Rota Rod Test with Male *Swiss Albino* (weight 25 g, 4 weeks) for 2 Minutes.**

## Open Field Test

The open field is a very popular animal model of anxiety-like behaviour that can also be used to detect extracts and phytoconstituents that elicits motor incoordination, CNS stimulation or depression. Following administration of vehicle, extracts, phytoconstituents or reference drugs, animals are placed in a circular or rectangular open field with known ðr2 or length x breadth, respectively, with a known height of wooden wall. A 5-min test session is often videotaped in a soundproof experimental room. Parameters such as crossing, rearing, grooming, defecation, urination, freezing activities are later scored. A reduction in rearing activity, total crossing and an increase in freezing could suggest sedative or myorelaxant effect at this dose. A reduction in the rearing or crossing activities in the open field could be associated to the habituation memory. An increase in crossing and time spent at the centre (effects that suggest a reduction in aversiveness and anxiety in this animal model) could suggest anxiolytic like property of extracts, phytoconstituents or reference drugs.

**Figure 7.5: Male *Swiss Albino* Mice Exploring Circular Open Field for 5 Minutes (Apparatus specification; Base area 62.80 cm² with a 50 cm high wooden wall.**

## 7.5 Evaluation of Putative Anxiolytic Properties

Contemporary biological psychiatry uses animal models to increase the understanding of affective disorder pathogenesis (Adam and Allan, 2014) and facilitate the discovery of new anxiolytic therapy. The spectrum of behavioural alterations through phenotypic domain in these models permits detection of extract or phytoconstituents with promising antianxiety. Animal models like open field, light dark box, elevated plus maze, social interaction, hole board test among others have been used extensively in the literature to evaluate anxiolytic like property of novel extract or phytoconstituents. The elevated plus-maze test, the light/dark test and the open-field test have been a mainstay of anxiolytic drug discovery research for many years. They assay anxiety-like behaviour by generating a conflict between a drive to approach novel areas and, simultaneously, to avoid potential threat therein. They have clear intuitive appeal, are inexpensive to construct, and ostensibly quick and easy to run (Griebel and Holmes, 2013).

Light–dark box test (LDB): The LDB is a test for behavioural phenotyping of mice and predicting anxiolytic or anxiogenic-like activity of drug. The light/dark transition test is one of the most widely used tests to measure anxiety-like behaviour in mice. The test is based on the natural aversion of mice to brightly illuminated areas and on their spontaneous exploratory behaviour in novel environments (Takao and Miyakawa, 2006). The apparatus consists of a dark chamber and a brightly illuminated chamber separated by a partition with door. According to Bourin and Hascoët (2003), a natural conflict situation occurs when an animal is exposed to an unfamiliar environment or novel objects. The conflict is between the tendency to explore and the initial tendency to avoid the unfamiliar (neophobia). Mice are allowed to move freely between the two chambers. Groups of mice are treated with vehicle, extract, phytoconstituents, reference drugs and placed (following a specific interval) at the centre of the light area facing the opening of dark area. The latency to the first transition, number of transitions between the two compartments and the time spent in the light area are recorded (Crawley and Goodwin, 1980). The exploratory activity reflects the combined result of these tendencies in novel situations. Thus, in the light/dark test, drug-induced increase in behaviours in the white part of a two-compartment box, in which a large white compartment is illuminated and a small black compartment is darkened, is suggested as an index of anxiolytic activity. An increase in transitions without an increase in spontaneous locomotion is considered to reflect anxiolytic activity.

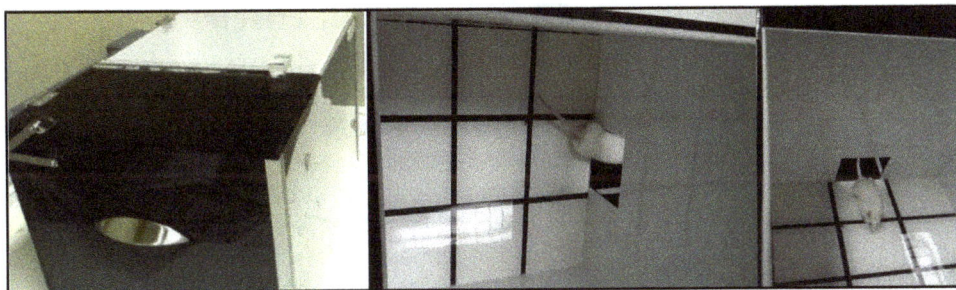

**Figure 7.6: Male *Swiss Albino* Mice Exploring Light Dark Box for 5 Minutes (Specification: 1/3ʳᵈ light and 2/3ʳᵈ dark area with an entrance or exit opening).**

## Elevated Plus Maze Test (EPM)

The elevated plus maze is a widely used behavioural assay for rodents and it has been validated to assess the anti-anxiety effects of pharmacological agents. The test is based on the natural aversion of mice for open and elevated areas, as well as on their natural spontaneous exploratory behaviour in novel environments. The apparatus consists of open arms and closed arms, crossed in the middle perpendicularly to each other, and a centre area (Komada *et al.*, 2008). Mice are given access to all of the arms and are allowed to move freely between them. Animals are placed at the junction of the four arms of the maze, facing an open arm. The number of entries and time spent at both the opened and enclosed arm are recorded by a video-tracking system following an interval between administration of extracts or phytoconstituents. Behaviours like rearing, head dips and stretched-attend postures can also be observed in this model (Walf and Frye, 2007). An increase in open arm activity (duration and/or entries) reflects anti-anxiety behavior. In our laboratory, rats or mice are exposed to the plus maze on one occasion; thus, results can be obtained in 5 min per rodent (Lister, 1987).

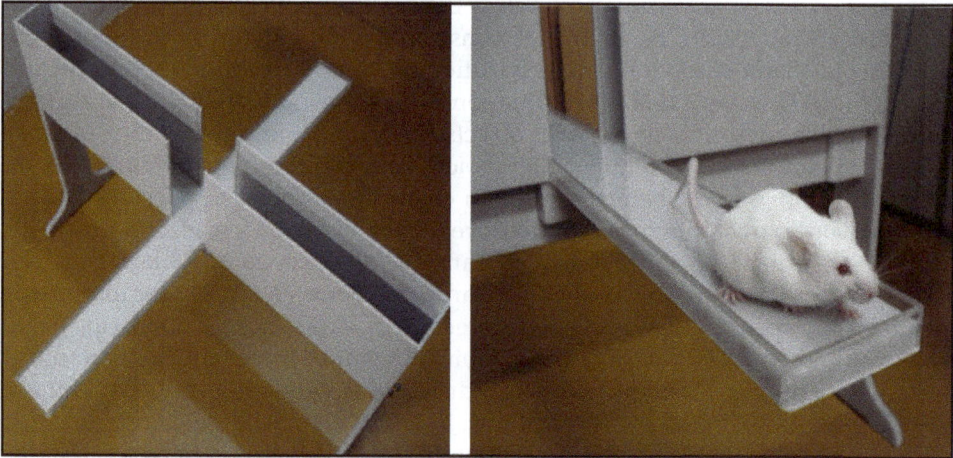

**Figure 7.7: Male *Swiss Albino* Mice Exploring Elevated Plus Maze.**

## 7.6 Evaluation of Putative Antidepressant Effects

Depression is a heterogeneous and multifaceted psychiatric disorder that poses enormous challenges to mimic and study in the laboratory. Despite the difficulties in translating the clinical affective disorders in its entirety into animal model, several innovative efforts which have been made over the years have advanced our knowledge about this diseases and facilitate drug development initiatives. Some of the progress, that have been made since the serendipitous discoveries of antidepressant can be attributed to the use of highly validated models, forced swimming test, tail suspension test sucrose consumption tests among others.

### Forced-Swimming Test (FST)

The forced swim test (FST), also known as the Porsolt test, is the most widely used animal model in depression research, more specifically, to screen putative antidepressant effect. The major advantage of the forced swim test is its relatively high throughput and ease of use. The test also provides a useful model in which to study the neurobiological and genetic mechanisms underlying stress and antidepressant responses (Lucki, 2001; Porsolt, 2000). FST, also known as behavioural despair test, is one of the most effective and widely acceptable preclinical models for the evaluation of antidepressant property of drugs. It was developed in the rat and later adapted to the mouse (Porsolt *et al.*, 1978, 1977a and b). The test is based on the observation that rodents, following initial escape-oriented driven movements, develop an immobile posture when placed in an inescapable water container. Besides active behaviours (climbing behaviour and swimming behaviour), measurement of immobility time is one of the most widely assessed parameter. Experimental animals are exposed to a glass cylinder of 15 cm diameter, 20 cm high filled with water to a height of 12 cm (for mice) or 30 cm diameter, 40 cm high filled with water to a height of 24 cm (for rats) with temperature of water was 24 °C ± 2°C. Animals are treated orally (chronically or acutely) with varying doses of extract/compound, reference drugs (imipramine or fluoxetine) or vehicle prior to the behavioural testing in FST. The test session is videotaped. At the end of animal exposure to FST, the container is cleaned with 10 per cent ethanol to prevent biasness that could emanate from odour from faeces and urine. Immobility time, is considered as a measure of behavioural despair to stressful condition. The time spent by the mouse floating in the water and making only those movements that were necessary to keep its head afloat is considered as immobility time. The shorter the immobility time the stronger the antidepressant effect of the extract, phytoconstituents under investigation (Cryan and Lucki, 2000). Measurement of other behaviours such as climbing and swimming in FST could suggest targets of antidepressant agents. False positives in the test include drugs that are stimulants (and hence decrease immobility) but are not antidepressant; these can be detected using complimentary open field test.

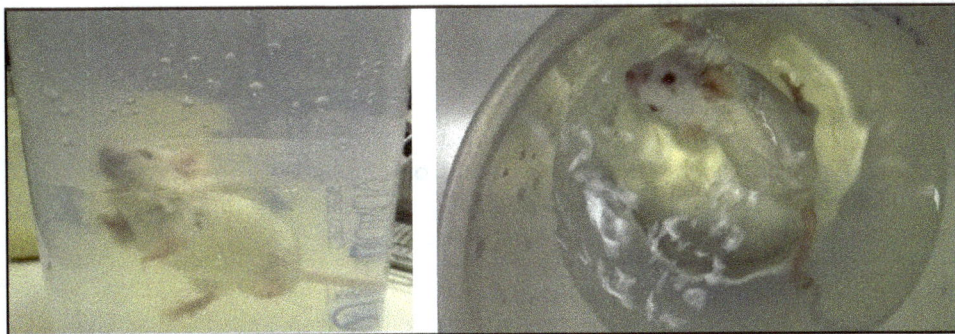

**Figure 7.8: Male *Swiss Albino* Mice Exploring during 6-minutes Forced Swimming Session (specification: Container of 42 cm in height with diameter of 18 cm, filled with water up to 30 cm at 24 ± 2°C).**

## Tail-Suspension Test

The tail suspension test was based on the method of Steru *et al.* (1985). The tail-suspension test is a mouse behavioural test useful in the screening of potential antidepressant drugs, and assessing of other manipulations that are expected to affect depression related behaviours. An inescapable stress associated with the suspension of a mouse by its tail could reveal the ability of animals to cope with a stressful situation. Antidepressant drugs or an extract or phytoconstituents with potential antidepressant property could enhance the ability to sustain a longer movements and reduce the period of passive posture or immobility time. Animals are treated orally with varying doses of extract, phytoconstituents, reference drugs or vehicle prior to the tail-suspension test. During the test, mice are suspended using an adhesive tape. Immobility time is measured when the mice assume a passive posture after the initial period of relentless movement.

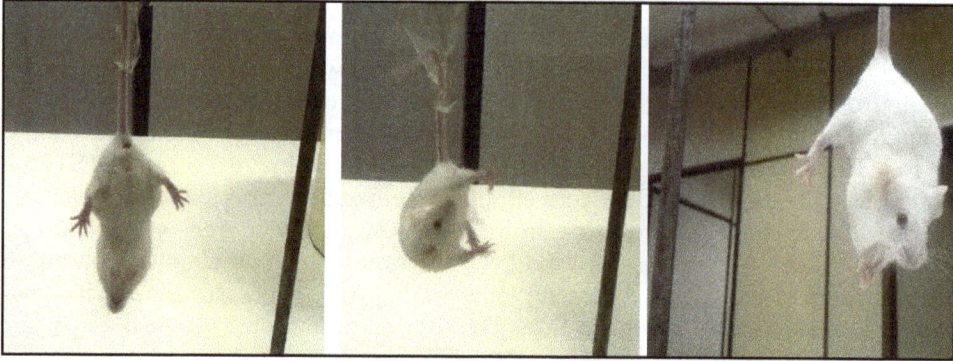

**Figure 7.9: Male *Swiss Albino* Mice during 6-minute Tail Suspension Test (Specification: adhesive tape that is placed 2 cm from the tip of the tail at about 50 cm above the floor).**

## Sucrose Consumption Test

Chronic stress-induced anhedonia. Loss of interest in pleasurable activities remains one of the common symptoms found in depressed patients. Hedonic deficits can be induced in rodents by chronic stress through varieties of stimuli such as electric shocks, immobilization, swimming in cold water and tail suspension (Katz, 1981; D'Aquila *et al.*, 1997; Willner *et al.*, 1987) Animals that are subjected to a regimen of chronic, mild, unpredictable stress exhibit behavioural deficits consistent with a loss of responsiveness to reward as reflected by a decrease in sucrose consumption. An agent with potential antidepressant activity could restore normal behaviour. The regimen of chronic stress could constitute the individual exposure of mice to rat (mice are introduced into rat cage of size 30 cm x 20 cm x 12 cm with a separating perforated walls that allows olfactory and visual contact of mice to rat and free access to water and food during 10 h). The exposure is followed by confinement stress and forced swimming in cold water. For the purpose of sucrose consumption test, mice are allowed

**Figure 7.10: Male *Swiss Albino* mice in home cage with free access to sucrose solution.**

free choice between a plastic recipient of 2 per cent sucrose solution and distilled water for 24 h. The estimate of water and sucrose consumption by control and experimental groups are assessed by measuring the recipients. The preference for sucrose was calculated as a percentage of consumed sucrose solution of the total amount of consumed liquid. A reduction of sucrose preference below 65 per cent, measured at 4 weeks of continuous stress application, was taken as the criterion for anhedonia (Strekalova *et al.*, 2004). This criterion was based on the fact that none of the control mice exhibited less than or equal to 65 per cent preference for sucrose. Based on this criterion, mice are assigned to non anhedonic or anhedonic groups prior to further testing or data analysis.

## 7.7 Evaluation of Putative Anticonvulsant Effects

The discovery and development of a novel anticonvulsant drugs depends on animal models, established efficacy and safety. Animal models for convulsion or seizures have advanced the discovery of novel anticonvulsant drugs and understanding of the mechanisms underlying ictogenesis. Despite the crucial role of animal models of seizures or epilepsy in drug discovery, some of the models been used to study ictogenesis may not be suitable for epileptogenesis. Some of the common animal models include pentylenetetrazol induced convulsion, seizures induced by maximum transcorneal electroshock, bicuculline, cholinomimetics, strychnine, insulin, penicinina, kainic acid, picrotoxin and pilocarpine among others.

### Pentylenetetrazol Induced Convulsion

The test of convulsion induced by pentylenetetrazol (PTZ) is a validated experimental model for preliminary evaluation of drugs with potential anticonvulsant activity. In this test, the latency to the onset of the first myoclonic and duration of the convulsive crisis are recorded. Other parameters like survival index and the severity of the seizure (a scale measure of collective changes in animal behaviour such as myoclonic jerks, vocalization, straub, akinesia, tremor, leap, paralysis, clonic seizure, rigidity and tonic extension of the hind limbs with death). Mice are randomly divided into groups with different oral treatments of extract/ compounds at varying doses, reference drug or vehicle prior to the intraperitoneal administration of pentylenetetrazol (PTZ). Animals are placed under camera to

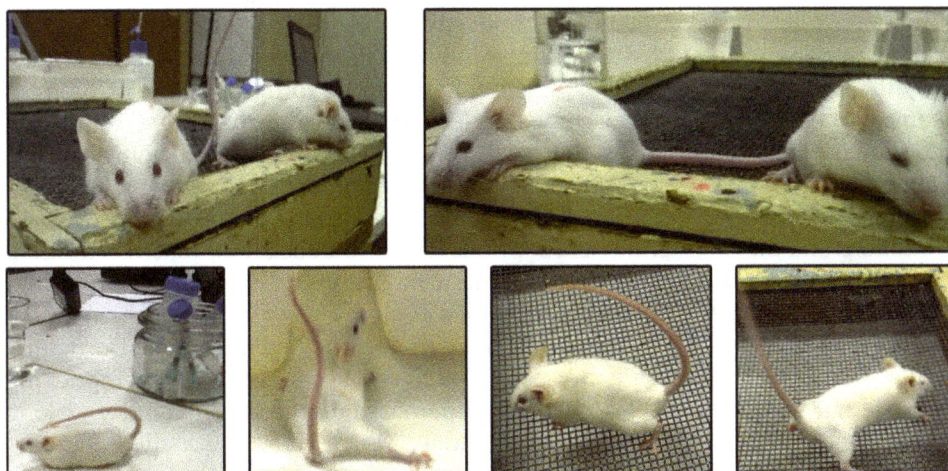

**Figure 7.11: Pentylenetetrazol Induced Convulsion in Male *Swiss Albino* Mice.**

record behavioural changes for 30 min. An increase in latency to the first myoclonic jerk, decrease in duration of seizure, attenuation of seizure severity or increase in survival index suggest anticonvulsant property.

### Seizures Induced by Maximum Transcorneal Electroshock

Seizures induced by maximum transcorneal electroshock has been used to investigate a possible anticonvulsant activity of novel extract/compounds. Mice are subjected to oral treatment with extract/compounds at varying doses, reference drugs (phenytoin, diazepam) or vehicle at appropriate interval prior to maximum transcorneal electroshock (60 Hz, 50 mA, 0.2 s). The animals are subsequently placed in the observation box to record behavioural alterations. Parameters like duration of hind paw extension and recovery can be evaluated. Extract or phytoconstituents that prevent the appearance of convulsions or block the extension of paws are capable of being used clinically to treat epilepsy.

## 7.8 Evaluation of Memory and Cognitive Deficits

Animal research on learning and memory could provide insight into the understanding of the neurobiology of cognition and the discovery of new therapy for the treatment of cognitive deficits. Understanding the neural basis of learning and memory is key to the resolution of many psychiatric diseases. For instance, depression and anxiety could be associated with experiences and aspirations of individual in a manner that the negative reinforcement of memory could keep an individual in a perpetual pathological state. The maladaptive automatic behaviours (drug-seeking behaviours) which are common among drug addicts are consequence of stable and organized memory. Behavioural changes are consequence of innate responses to stimulus. Stimulus, either pleasant or unpleasant experience, often leads to learning and memory formation. Memory formation progresses from labile to more stable forms as in habit. Habit formation involves sequential acquisition, repetition and motor behaviours elicited by external or internal triggers that at initiation progress to completion without constant conscious oversight (Graybiel, 2008). The context, quality, duration and intensity of the stimulus could determine the attributes of the memory or habits that are to be formed. Different types of memory have been identified clinically despite some controversies arising from the limitations been posed by the level of understanding of their neurobiology. Habit memory could be considered as implicit or non-declarative memory in that it involves gradual acquisition through multiple trials or exposure to learning conditions and the establishment of stimulus response associations that may develop outside of awareness, and are rigidly organized. In this case, experience modifies behaviour without any conscious memory content or even the experience that memory is being used. Non-declarative memory is expressed through performance, while declarative memory is expressed through recollection, as a way of modelling the external world. In addition, working memory, short term memory and long term memory among others have been widely discussed and modelled. Plethora of preclinical cognitive studies have shown that pharmacological interventions can exert promnesic (procognitive) and/or nootropic effects in laboratory animals.

Meanwhile, the fact that variability in behaviour among species, populations, or individuals has some genetic basis informed painstaking choice of mice, rats or their inbred strains for experimental purpose. Despite some limitations that are associated to animal models, some of the task in these models still offer great insight into the understanding of learning and memory. Learning can be assumed to have taken place through behavioural changes to environmental stimulus. Meanwhile, experimenter should pay attention to any interference with noise, light intensity, stress, odour, colour of apparatus, species life expectancy/age, intervals between training and test session. It must be stated that some non-traditional mnemonic models can be used to test memory. It is possible to test reference motor memory (long term, skill or procedural memory) in the rota rod and open field as described above. Some of the classical pharmacological approaches in use are as followed;

## Step-Down Inhibitory Avoidance

In this model, animals learn to inhibit their innate exploratory drive in a new environment (step-down apparatus) in order to avoid an unpleasant and non-lethal electric footshock (aversive stimulus). The inhibitory avoidance training apparatus (for rat) is a plastic or metal box (50 x 25 x 25 cm) with parallel 10-mm caliber bronze bars on its floor. The left end of the grid was occupied by a 7-cm wide, 2.5-cm high non-conductive platform. Mouse is generally placed on the platform and the latency to step down with four paws on the grid is automatically recorded during training and test sessions. The rats are treated orally or intraperitoneally with the vehicle, extract, compound or reference prior to training or post-training and a retention test of the same task is carried out after a specific interval (depending on the type of memory). In the training session, the mouse receives a footshock of varying intensities, for example 0.1 - 1 mA, after stepping down on the electrified grid until the animal escapes to the platform. The shock is expected to be delivered in few seconds (< 5s); if the animal fails to climb back onto the platform within 10 s it may be removed and excluded. Animals that steps down at time greater or equal to 20 s (this limit at times varies from laboratory/experimenter to laboratories/experimenter) are excluded from tests. The test session is performed after a defined interval (depending on the type of memory under scrutiny - short term or long term memory). The ceiling of step-down latency should be predetermined prior to the commencement of test session. The step-down latency (SDL) is a good parameter of learning and memory performance in this test. The SDL is taken during this session as a measure of retention (Salgueiro *et al.*, 1997). In order to validate this model and show that this experimental set up or task engender learning, control group is expected to exhibit unimpaired memory by demonstrating a higher SDL during training session than test session. Pharmacological treatment may not alter cognitive processes in this model or elicit either amnestic or procognitive effect. The absence of difference between the values of SDL during training and test sessions or if the value of SDL is lower than the control suggests amnesia. In the case of procognitive effect of pharmacological treatment the value of SDL is expected to either be higher during test session than training session or test measurements of experimental animal should be higher than the control-group.

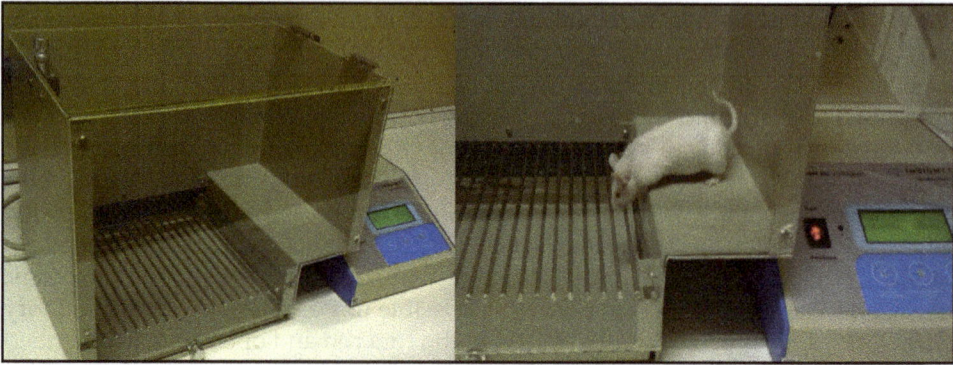

**Figure 7.12: Male *Swiss Albino* Mouse on Step Down Apparatus.**

## Object recognition task

Laboratory animals like rats and mice have a tendency to interact more with a novel object than with a familiar object. This tendency has been used by behaviourists to study learning and memory. An object-recognition task is a classical model that allows evaluation of the level of awareness, attention and matching accuracy of the animals to previous stimulus or experienced in a different context. Researcher should ensure animal's habituation to the experimental room and the apparatus in the absence of objects in dim light conditions prior to an acquisition phase in the presence of the object, retention interval and a test phase (preference evaluation phase). The duration of habituation, lightning, interval between acquisition and test sessions among others should be pre-established by the researcher. Animals are first placed in an apparatus (a plastic or wooden box) and allowed to explore two identical objects unfamiliar to mice (A1 and A2) within 5 min. Exploration is a measure of time spent sniffing or touching the object with the nose and/or forepaws. Animals failing to explore objects for at least 10 s are discarded from the study. After a prescribed interval, the animal is returned to the apparatus, which now contains the familiar object and a novel object. Object recognition is distinguished by more time spent interacting with the novel object. All objects presented similar textures, colours, and sizes, but distinctive shapes; after each trial, objects are washed with 10 per cent ethanol. A higher exploration time with novel object indicate higher capacity of the animals to perceive, discriminate, identify and recognize.

## Typical Experimental Procedure

In the first week, animals were acclimatized (twice for 3min each day) of the object recognition task apparatus (a circular arena with 83cm in diameter and half of the 40-cm-high wall is made of gray polyvinyl chloride, the other half of transparent polyvinyl chloride) without any objects. The light intensity (<"20 lx) was maintained in the whole apparatus. In the two following weeks, the rats were trained until they showed a stable discrimination performance. Subsequently, drug testing began. A testing session comprised two trials. The duration of each trial was 3 min. During the first trial (T1), the apparatus contained two identical objects (two objects that could not be displaced by a rat). A rat was always placed in the apparatus facing the wall at the middle of the front (transparent) segment. After the first exploration period, the rat was put back in its home cage. After a 1-h retention interval, the rat was returned to the apparatus for the second trial (T2), but now with dissimilar objects, a familiar one and a novel one. The time spent exploring each object during T1 and T2 was recorded manually, using a personal computer.

Exploration is defined as follows: directing the nose to the object at a distance of no more than 2 cm and/or touching the object with the nose. Sitting on the object was not considered as exploratory behaviour. In order to avoid the presence of olfactory trails, the objects were always thoroughly cleaned (with water and alcohol). Moreover, each object was available in triplicate so none of the two objects from the first trial had to be used as the familiar object in the second trial. In addition, all combinations and locations of objects were used in a balanced manner to reduce potential biases due to preferences for particular locations or objects.

**Figure 7.13: Male *Swiss Albino* Mouse Exposed to Object Recognition Task Apparatus (A rectangular arena with 80 x 60 cm and 40-cm-high transparent polyvinyl chloride wall.**

## Morris Water Maze

The Morris Water Maze Task is a widely validated behavioural task for the measurement of spatial memory of laboratory animals. In this model, animal learns to swim in a water tank and find a submerged platform in the presence of external cues (Morris, 1984). This model may be used to evaluate operant-like spatial learning. This task involves consciousness and associative capacity of the animals to environmental cues to the hidden platform. The time it takes the animals to find the escape platform and the tracing of animal's movement during trials are reference parameters that are often used to make inference about learning in this model. The colour, temperature, environmental cues and dimension of submerged platform and water maze should be adapted to the kind of experimental subject. The platform can be located at the centre of one of the quadrants or shifted to another on every trial during the training session. Alternatively, the experimenter can change the animals' starting points. Training sessions involves repeated exposure of the animals to water maze. During these sessions, platform position can be changed randomly or introduction of the animal into the model from different starting points. The escape latency and distance covered in locating the platform can be digitally recorded. Animals are withdrawn from the water maze, dried and kept in their home cage after each exposure prior to the next trial. The choice of inter-trial interval depends on factors like the kind of memory attribute under investigation, species, etc. The test session can start after a predetermined time interval that is meant for memory consolidation phase following the training session. The latency to locate the original position of the platform for the first time, distance covered, number of crossings and time spent in the quadrant where the platform is located are important parameters to be evaluated during test session. The animals in the control group is expected to demonstrate smaller test session escape latency and cover shorter distance during test session than training session for normal memory retention. These behavioural changes during the test session indicate that the animal has learned to escape from the stressful swimming task. In addition, the behaviour of control group is essential in determining reliability of the set-up in the evaluation of memory and if the experimenter is capable of executing this model. In this group, the number of crossing and the time spent in the quadrant with platform are expected to be higher during test session than training session. Other groups could be proposed to verify memory impairment, amnestic or prominesic effects of extracts, compounds or reference drugs. Some tests are designed to evaluate if extracts or compounds are capable of reversing/attenuating memory impairment induced by lesions (surgical or chemical), electrical stimulations, environmental or genetic manipulations. An amnestic treatment could be suggested if group of animals did not show difference in the escape latency between training and the test sessions could suggest. A lower escape latency of test session in the experimental group as compared to the value during the test session in control group could also

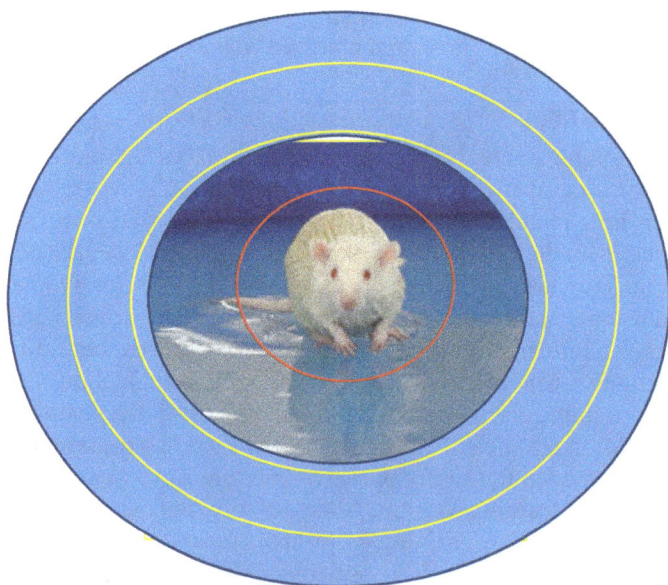

**Figure 7.14: Representation of Mouse on the
Centrally Placed Escape Platform in Water Maze.**

be referred to as amnesia. In contrary, a facilitatory treatment can be suggested when the escape latency of experimental group is higher during training sessions than test sessions or escape latency is smaller during test session than the control-group.

## ARMS Radial Maze

The Arms Radial Maze is a wooden or plastic apparatus with characteristic number of arms (4, 6, 8, 12) constructed radially around a central arena or platform that is elevated from the floor. Arms of varying dimensions (length and breadth) may be devoid of elevated walls. Sometimes guillotine doors, floors with distinct textures or illumination inside arms are used as part of environmental context to reinforce learning. A small translucent plastic dish (a food recipient) of known dimension is sometimes mounted at the end of each arm for baiting. This food recipient may not be visible to the animals. The animals are placed on a restricted diet to maintain them at about 85 per cent of free-feeding weight. This procedure could serve as a drive to execute the task of locating bait. The baiting pattern of arm often remain constant during training so as to allow the learning of the spatial pattern of baited and unbaited arms. Training sessions may begin with or without prior habituation of the animal to the apparatus through one or more exposure to the Arms Radial Maze per day or last for days or weeks in a consecutive or alternate manner. The animals are allowed to find the baits. Evidence that they did so was obtained by testing the rats in the same maze. The entry of the animal (with the crossing of four paws into an arm) previously entered in the same session may also be considered an error that suggests a working memory deficit. Following a predetermined period of delay interval, animals are exposed to the maze for retention test. The arms that were entered and the sequence of entry are recorded for further analysis. Pre-training, post-training or pre-test administration of extracts or compounds could affect spatial learning acquisition or memory retrieval. The use of non-mnemonic or complementary task to verify effect of drugs on memory is essential. Pharmacological treatment that impaired or decreased the visual capacity of animals to see cues in the Morris Water Maze Task could mask learning behaviour and lead to false result. An anxiolytic drug that lessened the animals' anxiety in the water could decreas the desire to escape. In addition, an agent that intereferes with locomotion activity of the animals could mask the spatial memory of the animals.

## Chapter 8

# Overview of Medicinal Plants with Cardiovascular Activities

*James O. Fajemiroye, Lara M. Naves*
*and Gustavo R. Pedrino*

## ABSTRACT

*Cardiovascular disease, constitute serious health challenges and cause of death worldwide. Meanwhile, medicinal plants have been used for the treatments of cardiovascular diseases for millennia. This may be associated with their ability to act as antioxidants, vasodilators or calcium, adrenoceptor and platelet activating factor (PAF) antagonists. The cardiovascular activities of some of these plants have been explored as a therapy in patients with hypertension, hyperlipidemia, thromboembolism, coronary heart disease, congestive heart failure, angina pectoris, atherosclerosis, cerebral insufficiency, venous insufficiency, arrhythmia, etc. Although, preclinical assessment of these plants supported their therapeutic potentials, the safety and interactions with other drugs remain sources of concern. Considering the increase in the number of patients with cardiovascular diseases, it becomes more interesting to be aware of the wide array of medicinal plants that are available and their respective beneficial and adverse effects. Hence, there is a need for careful consideration of potential benefit and adverse effects of medicinal plants along with the clinical history of the patients prior to application. Considering toxicity profile and low therapeutic index of many cardioactive species, arbitrary use of these medicinal plants should be avoided. In this chapter, some of the cardioactive plant species and active principles such as tetrandrine, berbamine, neferine, dauricine, stephanine, dicentrine, berberine, tetrahydropalmatine, stepholidine, tetramethylpyrazine rhynchophylline, hirsutine, ginkgolides, praeruptorins, scoparone, ginsenosides, dehydroevodiamine, dictamine and fraxinellone, emodin, magnolol, norathyriol, oxymatrine, sophoramine, paeonol, puerarins, rhomotoxin, schisanhenol, sinomenine and tussilagone are listed along with available pharmacological data and animal models.*

## 8.1 Introduction

Cardiovascular diseases are currently the leading cause of death in industrialized countries (Baharvand-Ahmadi, Bahmani, and Zargaran, 2016). Hypertension, hyperlipidemia, thromboembolism, coronary heart disease and heart failure are among broad range of cardiovascular diseases (Khosravi-Boroujeni *et al.*, 2013; Sadeghi *et al.*, 2014). Hypertension is known as a risk factor for the diagnosis of myocardial infarction, stroke and peripheral vascular disease (Asgary *et al.*, 2014; Baradaran, Nasri, and Rafieian-Kopaei, 2014).

Medicinal plants have always been valuable natural resources as a result of their culinary and medicinal properties. A wide variety of extracts from these plants have been used for years as alternative source of medical care mostly in the developing world. The historical background of plants based drug including ephedrine from *Ephedra sinica*, digitoxin from *Digitalis purpurea, salicin* (the source of aspirin) from *Salix alba*, and reserpine from *Rauwolfia serpentine* (Mashour *et al.*, 1998) are evidences of medicinal plants importance towards drug development. These plants contains active principles which in turn possess highly diverse and complex molecular structures. One of the problems among the patients and physicians has been the paucity of scientific data on medicinal plants in circulation (Mashour *et al.*, 1998). This chapter listed some of the medicinal plants and active principles with cardiovascular activities. The animal models that are being used for preclinical investigation of these cardioactives are also highlighted.

## 8.2 Some Species with Cardiovascular Activities

### *Acanthospermum hispidum* DC

*Acanthospermum hispidum,* commonly known as the Bristly starbur, Goat's head, is an erect annual herb that belongs to the family Asteraceae; and originated from tropical America (Chakraborty, Gaikwad, and Singh, 2012). Araújo *et al.* (2008) reported that the plant has been used traditionally for the treatment of hypertension. Schaeffer *et al.* (2012) stated that several studies have demonstrated that flavonoids can significantly lower systolic blood pressure.

### *Digitalis* spp.

*Digitalis* spp. contain potent cardioactive glycosides, which have positive inotropic actions on the heart. These glycosides have a low therapeutic index. The drugs digitoxin, derived from either *D. purpurea* (foxglove) or *Digitalis lanata*, quinine from *Cinchona ledgeriana, digoxin*, derived from *D. lanata* alone, have been used in the treatment of congestive heart failure for many decades (Mashour *et al.*, 1998).

### *Rauwolfia serpentina*

The root of *Rauwolfia serpentine* is the natural source of the alkaloid reserpine. The use of *Rauwolfia serpentina* root for the treatment of hypertension and psychoses was first described in 1931. Its isolate, reserpine, has been subsequently used on a large scale to treat systemic hypertension. This compound irreversibly blocks the uptake of monoamines. The sympatholytic and antihypertensive actions of reserpine has been associated with the depletion of catecholamines. Reserpine lowers blood pressure by decreasing cardiac output, peripheral vascular resistance, heart rate, and renin secretion (Mashour *et al.*, 1998).

## Stephania tetrandra

*Stephania tetrandra* is used in traditional Chinese medicine to treat hypertension (Mashour *et al.*, 1998). Tetrandrine, an alkaloid isolated from *S. tetrandra* extract, has been described as an antagonist of calcium ion channel since it blocks T and L calcium channels. It also interferes with the binding of diltiazem and methoxyverapamil at calcium-channel binding sites, and suppresses aldosterone production (Rossier *et al.*, 1993; Sutter and Wang, 1993). An oral dose of 25 or 50 mg/kg induced a gradual and sustained hypotensive effect after 48 hours in a stroke-prone hypertensive rats (Kawashima *et al.*, 1990). Tetrandrine orally administered 3 times weekly for 2 months in dogs also induced liver necrosis at 40 mg/kg and reversible swelling of liver cells at 20-mg/kg (Sutter and Wang, 1993).

## Lingusticum wallichii

The root of *Lingusticum wallichii* is used in traditional Chinese medicine as a circulatory stimulant and hypotensive drug (Ody, 1993). Tetramethylpyrazine, the active constituent extracted from *L. wallichii*, inhibits platelet aggregation *in vitro* and lowers blood pressure by vasodilation in dogs. With its actions independent of the endothelium, tetramethylpyrazine's vasodilatory effect is mediated by calcium channel antagonism and nonselective antagonism of α-adrenergic receptors.

## Uncaria rhynchophylla

*Uncaria rhynchophylla* is sometimes used to treat hypertension (Mashour *et al.*, 1998). Its indole alkaloids, rhynchophylline and hirsutine, are thought to be the active principles that are responsible for the vasodilatory effect. *In vitro* study showed that *U. rhynchophylla* extract relaxed norepinephrine-precontracted rat aorta through endothelium-dependent and -independent mechanisms. *U. rhynchophylla* extract seems to stimulate endothelium-derived relaxing factor and/or nitric oxide release for endothelium-dependent component (Kuramochi, Chu, and Suga, 1994).

## Veratrum spp.

*Veratrum* spp. which include *Veratrum viride, Veratrum californicum, Veratrum album* and *Veratrum japonicum*, is a perennial plant that grows worldwide. *Veratrum spp* contain poisonous alkaloids that causes vomiting, bradycardia, and hypotension. Its application in the treatment of hypertension has been discouraged due to its low therapeutic index and unacceptable toxicity. The availability of safer antihypertensive drug makes its therapeutic demand (Jaffe, Gephardt, and Courtemanche, 1990). Alkaloids from *Veratrum genus* enhance nerve and muscle excitability by increasing sodium ion conductivity. They act on the posterior wall of the left ventricle and the coronary sinus baroreceptors, causing reflex hypotension and bradycardia via the vagus nerve (Mashour *et al.*, 1998).

## Evodia rutaecarpa

*Evodia rutaecarpa* is a Chinese herbal drug that has been used as a treatment for hypertension (Mashour *et al.*, 1998). It contains an active vasorelaxant component called rutaecarpine that can cause endothelium-dependent vasodilation in experimental models (Chiou *et al.*, 1997).

## Crataegus hawthorn

*Crataegus hawthorn* encompasses many Crataegus species such as *Crataegus oxyacantha, Crataegus monogyna, Crataegus pinnatifida.* These are considered as important tonic for the cardiovascular system. It is particularly useful for angina (Mashour *et al.*, 1998). *Crataegus* extract antagonizes the increases in cholesterol, triglyceride, and phospholipid levels in low-density lipoprotein (LDL) and very low-density lipoprotein in rats fed a hyperlipidemic diet. Hence, it may inhibit the progression of atherosclerosis (Shanthi, Parasakthy, Deepalakshmi, and Devaraj, 1994). This hypocholesterolemic action may be due to an up-regulation of hepatic LDL receptors that causes greater influx of plasma cholesterol into the liver. *Crataegus* also prevents cholesterol accumulation in the liver by enhancing cholesterol degradation to bile acids, as well as suppressing cholesterol biosynthesis (Rajendran *et al.*, 1996). In addition, high concentration of *Crataegus* extract has a cardioprotective effect on ischemic-reperfused hearts (Nasa *et al.*, 1993). The mechanism of *Crataegus* species cardiac action may be due to the inhibition of the 3′, 5′-cyclic adenosine monophosphate phosphodiesterase (Schüssler, Hölzl, and Fricke, 1995). In comparison with other inotropic drugs such as epinephrine, amrinone, milrinone, and digoxin, *Crataegus* reduced potentially arrhythmogenic risk due to its ability to prolong the effective refractory period (Joseph, Zhao, and Klaus, 1995; Pöpping *et al.*, 1995).

## Salvia miltiorrhiza

*Salvia miltiorrhiza*, a relative of the Western sage *Salvia officinalis*, is native to China. In traditional Chinese medicine, the root of *S. miltiorrhiza* is used as a circulatory stimulant, sedative, and cooling drug (Ody, 1993). Since *Salvia miltiorrhiza* could dilate coronary arteries in all concentrations, it may be useful as an antianginal drug. *S. miltiorrhiza* inhibits platelet aggregation and serotonin release induced by either adenosine diphosphate or epinephrine. This activity seems to be mediated by S miltiorrhiza's inhibition of cyclic adenosine monophosphate phosphodiesterase that causes an increase in platelet cyclic adenosine monophosphate (Wang *et al.*, 1982). *Salvia miltiorrhiza* seems to have a protective action on ischemic myocardium by enhancing the recovery of contractile force (Yagi, Fujimoto, Tanonaka, Hirai, and Takeo, 1989). A decoction of *S. miltiorrhiza* was as efficacious as the isolated tanshinones (Zhang, Wojta, and Binder, 1994).

## Allium sativum

*Allium sativum* (Garlic) has been valued for centuries for its medicinal properties. Intact cells of garlic bulbs include an odorless, sulfur-containing amino acid known as allinin. The allinin in garlic crushed is converted to allicin by allinase. Fresh garlic releases allicin in the mouth during the chewing process (Mashour *et al.*, 1998). Allicin and its derivatives are believed to be the active constituents of garlic's physiological activity. Many researches have focused on garlic's use in preventing atherosclerosis. Garlic has demonstrated multiple beneficial cardiovascular effects which include lowering blood pressure, inhibiting platelet aggregation, enhancing fibrinolytic activity, reducing serum cholesterol and triglyceride levels, and protecting the elastic properties of the aorta (Kleijnen, Knipschild, and Riet, 1989).

The precise extent of garlic's impact on atherosclerosis remains controversial. Long-term garlic powder intake may have a protective effect on the elastic properties of the aorta related to aging (Breithaupt-Grögler *et al.*, 1997).

## Ginkgo biloba

*Ginkgo biloba*, a species that was apparently saved from extinction by human intervention. The flavonoids from *Ginkgo biloba* reduce capillary permeability as well as fragility and serve as free radical scavengers. The ginkgolides (terpenes) inhibit platelet-activating factor, decrease vascular resistance, and improve circulatory flow without appreciably affecting blood pressure (Mowrey, 1993; Z'Brun, 1995). Experimental data seem to support the primary use of *G. biloba* for treating cerebral insufficiency. *G. biloba* appears to be useful for treating peripheral vascular disease (Doly, Droy-Lefaix, and Braquet, 1992; Kleijnen and Knipschild, 1992; Mowrey, 1993; Samson, Ramachandran, and Le Jemtel, 2014; Tyler, 1994; Z'Brun, 1995). Standardized extract of *G. biloba* with respect to its flavonol glycoside and terpene lactone content (EGb 761) seems to be valuable in the treatment of peripheral artery disease (Mouren, Caillard, and Schwartz, 1994).

## Rosmarinus officinalis

*Rosmarinus officinalis* (rosemary) known mostly as a culinary spice and flavoring agent has been listed as a tonic and stimulant. Traditionally, rosemary leaves are said to enhance circulation, aid digestion, elevate mood, and boost energy. The flavonoid pigment diosmin is reported to decrease capillary permeability and fragility (Swain, Dutton, and Truswell, 1985). The therapeutic use of rosemary for cardiovascular disorders remains questionable as only few clinical trials have been conducted on this species. Despite the cutaneous vasodilation as counterirritant properties of rosemary's essential oils, no evidence to support any prolonged improvement in peripheral circulation (Tyler, 1994).

## Aesculus hippocastanum

The seeds of *Aesculus hippocastanum* (horse chestnut) have long been used in Europe to treat varicose veins. *Aescin* (saponin glycoside) from the extract of this species inhibits the activity of lysosomal enzymes. Lysosomal enzymes contribute to varicose veins by weakening vessel walls and increasing permeability which subsequently result in dilated veins and edema (Tyler, 1994). *Aesculus hippocastanum* extract, in a dose-dependent manner, increases venous tone, venous flow, and lymphatic flow in experimental animals. This extract also antagonizes capillary hyperpermeability induced by histamine, serotonin, or chloroform. The suppression of experimentally induced pleurisy and peritonitis through inhibiting plasma extravasation and leukocyte migration by *A. hippocastanum* extract showed its anti-exudative properties. The extract of this species also improved symptoms that are associated with chronic venous insufficiency, such as pain, tiredness, itching, and tension in the swollen leg (Greeske and Pohlmann, 1996). In addition to its effects on venous insufficiency, prophylactic use of this extract seems to decrease the incidence of thromboembolic complications of gynecological surgery (Mashour *et al.*, 1998).

## Ruscus aculeatus

*Ruscus aculeatus,* commonly known as butcher's broom, has been used in treating venous insufficiency. *R. aculeatus* is a short evergreen shrub found commonly in the Mediterranean region. Ruscogenin and neurogenin (saponins) that are isolated from the rhizomes of *R. aculeatus* are thought to be its active components (Tyler, 1993). *In vivo* studies on hamster cheek pouch revealed that topical extract of this species antagonized histamine-induced increases in vascular permeability (Bouskela, Cyrino, and Marcelon, 1994a). In addition, *Ruscus* extract causes dose-dependent constriction of venules without a significant alteration on arterioles (Bouskela, Cyrino, and Marcelon, 1994b). Based on the data from the administration of prazosin, diltiazem, and rauwolscine, the peripheral vascular effects of Ruscus extract appear to be selectively mediated by effects on calcium channels and $\alpha$1-adrenergic receptors (Bouskela *et al.*, 1994a, 1994b). Although there are many evidences to support the pharmacological activity of *R. aculeatus*, there is still a relative few clinical data to establish its safety and efficacy.

# 8.3 Biological Activities of some Isolated Cardioactives

## Tetrandrine

Tetrandrine, an isolate of *Stephania tetrandru,* is a bis-benzylisoquinoline – consists of two benzylisoquinoline joined by an ether linkage. It induced hypotensive in man (Pharmacology, 1979). It has been shown to be a calcium antagonist (Fang *et al.*, 1981). Tetrandrine was found to completely inhibit diltiazem binding and increase nitrendipine binding, presumably through benzothiazepine sites and allosterically at other calcium associated sites (King *et al.*, 1988). Tetrandrine 15 mg/kg administered intravenously lowers mean, systolic, and diastolic blood pressure and depresses cardiac contractility for more than 30 minutes in conscious rats.

## Neferine

Neferine which shares structural similarities with tetrandrine is isolated from the seeds of *Nelumbo nucifera* (Li, Qian, and Lü, 1990). It seems to have quinidine-like activity on the heart and acts as calcium antagonist.

## Dauricine

Dauricine is an isolate of *Menispermum duuricum* (Lu and Liu, 1990). It acts as calcium antagonists (Lu and Liu, 1990). *In vivo* assay (anaesthetised rabbits) showed dauricine induced prolongation of A-H and H-V intervals and widening of ventricular action potentials in hearts (Zhu, Zeng, and Hu, 1990).

## Dicentrine

Dicentrine isolated from *Linderu megaphylla* (Teng *et al.*, 1990) is an aporphine. Dicentrine is reported to be $\alpha$1 competitive adrenoceptor antagonist. Dicentrine acts independently of the endothelium. At 30 µM of dicentrine, contraction induced by the thromboxane receptor agonist U-46619, angiotensin 11, high potassium, or carbachol remained unaltered, thereby suggesting that it is a potent and selective $\alpha$1 adrenoceptor antagonist in the aorta (Teng *et al.*, 1990).

# Berberine

Berberine is commonly found in species, especially *Stephania japonica,* belonging to berberidaceae. Berberine lowers blood pressure in animals. Although there is still controversy as to the mechanisms responsible, it induces positive inotropic effect on the heart in conscious rats, lowers peripheral resistance and slows the heart rate (Fang *et al.,* 1987).

# Tetrahydropalmatine

Tetrahydropalmatine (TH) has been isolated from *Corydalis ambigua.* TH has been reported to produce positive prior to negative, inotropic and chronotropic effects on the hearts of anaesthetised dogs (Liu and Zhao, 1987). The derivative of TH, benzyltrtrahydropalmatine - BTH considered as a potassium channel blocker (Yao *et al.,* 1990), induced antiarrhythmic actions in arrhythmias produced by coronary occlusion and reperfusion in cats (Xia *et al.,* 1990). BTH prolonged the duration of action potentials in guinea pig papillary muscles (Yao *et al.,* 1990). BTH competitively antagonised phenylephrine contraction rabbit aortic muscles and non-competitively reduced isoprenaline and histamine induced chronotropic effects (Chen *et al.,* 1987).

# Tetramethylpyrazine

Tetramethylpyrazine (TM) is an isolate of *Ligusticum wullichii* Franchat (Kwan, Daniel, and Chen, 1990). The extracts of *L. wullichii* have been used to treat vascular disease (Jiangsu New Medical College, 1979). TM inhibited platelet aggregation in vitro, induced vasodilatation and subsequently lowered blood pressure in anaesthetised dogs (Nie, Xie, and Lin, 1985). TMP and nitroprusside induced endothelium independent vascular activities (Liu *et al.,* 1990). The arterial dilating effects of TMP was not blocked by $N^G$-nitro-L-arginine methyl ester (L-NAME), an inhibitor of nitric oxide synthase.

# Rhynchophylline and Hirsutine

These indole alkaloids have been detected in *Uncaria rhynchophylla.* The extracts of this species have been used to treat hypertension (Jiangsu New Medical College, 1979). In vivo assay has shown vasodilatation effect of both rhynchophylline and hirsutine (Ozaki, 1990; Shi *et al.,* 1992). In aortic strips, hirsutine reduced intracellular calcium while rhynchophylline interfered with rabbit platelet aggregation and reduced thromboses induced in rats (Chen *et al.,* 1992). Both compounds are considered to alter calcium influx in response to activation of voltage sensitive channels.

# Ginkgolides

Ginkgolides are terpenoids naturally occurring in *Ginkgo biloba.* At least five ginkgolides have been identified (A, B, C, J, and M). Ginkgolide B, commonly referred to as BN 52021, is an antagonist of platelet activating factor (Braquet and Hosford, 1991). Platelet activating factor (PAF) has complex activities on the cardiovascular system. *Ginkgolide B* selectively inhibit PAF induced increased vascular permeability.

## Kadsurenone and Denudatin B

Both kadsurenone and denudatin B are PAF antagonists isolated from *Piper futokadsura* and *Magnolia fargesii*, respectively. Both kadsurenone and denudatin B inhibit KCl induced contraction of rat aorta and seem to act in part as calcium antagonists (Kecskeméti and Braquet, 1990; Yu *et al.*, 1990). Other compounds such as wallichinine, hancinone with PAF inhibition have been isolated from *Piper wallichii Hand.* and *Piper hancei Maxim* (Han *et al.*, 1989).

## Praeruptorins C

Praeruptorins C is a coumarin isolated from *Peucedanum praeruptorin* Dunn. The blood pressure of conscious rats was reduced by oral administration of praeruptorin C at the dose of 2 mg/kg. The vertebral resistance, coronary and femoral vessels were reduced in anaesthetised dogs at the doses of 20 and 100 µg/kg (Rao, Shen, and Zou, 1988). Praeruptorin C acts like a calcium antagonist by inhibiting (noncompetitively) positive chronotropic effects of isoprenaline on guinea pig atrium (Wu and Rao, 1990).

## Ginsenosides

Ginsenosides are complex glycosides found in *Panux ginseng CA Meyer*. Study have shown that ginsenoside $R_0$ and $R_b$ depressed lactic dehydrogenase release from neonatal rat myocytes during anoxia and reoxygenation. These compounds reduced creatine phosphokinase release in the isolated rat heart during reperfusion (Li, Deng, and Chen, 1987).

## Dehydroevodiamine

Dehydroevodiamine (DeHE) is a quinazolinocarboline alkaloid isolated from the fruit of *Evodiae rutaecarpa* (Loh *et al.*, 1992). The extracts of E. rutaecarpa are used to treat cardiovascular disease in traditional Chinese medicine (Jiangsu New Medical College, 1979). DeHE seems to possess quinidine- like action as it lowers blood pressure with negative inotropic. This compound is assumed to interact with sodium channels as well as potassium channels in the heart (Loh *et al.*, 1992).

## Dictamine and Fraxinellone

Dictamine and fraxinellone are isolated from *Dictamus dasycarpus*. The experimental result on rat aortic strips suggested that these compounds are calcium antagonists. The dictamine was suggested to act on voltage sensitive and receptor operated calcium channels while fraxinellone acts only on voltage sensitive ones (Yu *et al.*, 1990).

## Emodin

Emodin is an anthraquinone and a constituent of *Cascara sagrada, Rheum officianale* and *Polyganum multiflorum*, etc (Huei-Chen, Chai-Rong, Pei-Dawn Lee Chao, Ching-Chow, and Shu-Hsun, 1991; Jiangsu New Medical College, 1979). Emodin is a non-specific vasorelaxant. This relaxation can be inhibited or potentiated by free radical scavengers and quinacrine, respectively. Cyclic GMP levels were raised in the presence of emodin.

## Magnolol

Magnolol has been isolated from *Magnolia officinalis* and in rat aortic strips it causes rapid and slow relaxation of noradrenaline induced contractions. The removal of the endothelium abolished fast relaxation (Teng *et al.*, 1990). It was concluded that magnolol releases endothelium derived relaxing tactor and inhibits calcium influx through voltage sensitive channels (Teng *et al.*, 1990).

## Norathyriol

Norathyriol, a xanthone isolated from *Triptospermum lanceolatum*, belongs to the family of Gentianaceae. It seems to be d calcium antagonist since it inhibits KCI or CaCl induced endothelium independent contraction of isolated rat aorta (Ko *et al.*, 1991).

## Oxymatrine and Sophoramine

Oxymatrine and Sophoramine have been detected in *Sophora alopecuroides* and *Sophora flavescens*, respectively (Bian and Toda, 1988; Qin, Den, and Zhuang, 1990). Sophoramine is a non-specific inhibitor of prostaglandin F2α induced contraction of isolated arteries. Sophoramine showed antiarrhythmic actions in experimental animals and antagonised non-competitively the positive chronotropic actions of isoprenaline in rabbit atria (Li and Zhang, 1989). It does not antagonise contraction of rabbit aortic strips induced by KCl or noradrenaline (Yao and Zhang, 1989).

## Paeonol

Paeonol is an acetophenone from *Paeonia suffruticosa*. Paeonol inhibited calcium uptake in cultured heart cells from neonatal rats, however, the effect of this compound is about eight times less potent than verapamil in model (Tang and Shi, 1991).

## Rhomotoxin

Rhomotoxin has been detected in *Rhododendron molle* (Jin *et al.*, 1985). It has been used as an antihypertensive agent (Chen *et al.*, 1987). In isolated guinea pig papillary muscles, rhomotoxin induced positive inotropic effect, reduced resting potential, amplitude and duration of the action potential. Since tetrodotoxin antagonized this effect, it was concluded that rhomotoxin acts as sodium channel opener (Jin *et al.*, 1985).

## Sinomenine

Sinomenine occurs in *Sinonzenium acutum* (Thunb.). This compound has been used as an antihypertensive and antiarrhythmic agent (Jiangsu New Medical College, 1979). Sinomenine induced negative inotropic, decrease in upstroke velocity and amplitude of the action potential in isolated guinea pig papillary muscles (Li, Zhao, and Li, 1987). It was concluded that sinomenine blocks (non-selectively) sodium, calcium and potassium channels.

## Cardiac Glycosides

Cardiac glycosides comprise a large family of naturally derived compounds. Glycosides are found in many species including *D. purpurea, Adonis microcarpa, Adonis vernalis, Apocynum cannabinum, Asclepia scurassavica, Asclepias friticosa, Calotropis precera, Carissa spectabilis, Cerebra manghas, Cheiranthus cheiri, Convallaria majalis, Cryptostegia grandiflora, Helleborus niger, Helleborus viridus, Nerium oleander, Plumeria rubra, Selenicerus grandiflorus, Strophanthus hispidus, Strophanthus kombe, Thevetia peruviana* and *Urginea maritime* among others. Although, glycosides show considerable structural diversity, all members of this family share a common structural motif (Prassas and Diamandis, 2008). Cardiac glycosides have a long history of therapeutic application. The early understanding of their positive inotropic effects facilitated their use as effective drugs for the treatment of heart-related pathologies. Glycosides bind to and inhibit Na+/K+-ATPase. Members of this family have been in clinical use for many years for the treatment of heart failure and atrial arrhythmia, and the mechanism of their positive inotropic effect is well characterized (Prassas and Diamandis, 2008).

## 8.4 Intoxication and Adverse Effects of Cardioactive Plant or Active Principles

The fact that many medicinal plants have been in use for centuries, makes people have little or no doubt about its safety. The toxic effects that usually include life-threatening ventricular tachyarrhythmias, bradycardia, heart block and renal failure which could cause morbidity and mortality are associated with cardiovascular active medicinal plants and active principles. The outbreak of rapidly progressive renal failure after using a combination of several herbs as part of a dieting regimen is an indication of deadly consequence of improper utilization of cardioactive plants (Mashour *et al.*, 1998). Accidental poisonings have been associated with ingestion of cardiac glycosides (Bagrov *et al.*, 1995; Cheung, Hinds, and Duffy, 1989; Dickstein and Kunkel, 1980; Galey *et al.*, 1996; Langford and Boor, 1996; Moxley *et al.*, 1989; Nishioka el al., 1986; Safadi *et al.*, 1995; Shaw and Pearn, 1979). In 1993, 2388 toxic exposures in the United States were reported to be due to plant glycosides. Out of these number, 25 per cent (the largest percentage) was attributed to Oleander exposure (Litovitz, Clark, Soloway, 1993). For instance, all the oleander tissues, including the seeds, roots, stems, leaves, berries, and blossoms, are considered extremely toxic (Safadi *et al.*, 1995). Despite the therapeutic benefit of Digoxin (from *Digitalis* spp) in individuals with heart failure and/or atrial fibrillation (Gheorghiade *et al.*, 2006), digoxin toxicity may occur in a short period of time. Patients that are hypersensitive to *Rauwolfia alkaloid* such as reserpine could be susceptible to mental depression with suicidal tendencies. The most common adverse effects of *Rauwolfia alkaloids* are sedation and inability to concentrate and perform complex tasks. Reserpine's sympatholytic effect and its enhancement of parasympathetic actions account for its well-described adverse effects: nasal congestion, increased gastric secretion, and mild diarrhea (Oates *et al.*, 1996; Pharmacopeial Convention, 1998; Swain *et al.*, 1985; Tyler and Tyler, 1992). Clinically, when *S. miltiorrhiza* and warfarin sodium are coadministered, there is an increased incidence in warfarin-related adverse effects; in rats *S. miltiorrhiza* was shown to increase the plasma concentrations of warfarin as well as the prothrombin time (Chan *et al.*, 1995). In addition, allergic reactions to garlic have been reported (Delaney and Donnelly, 1996). Garlic should be used with caution in people taking oral anticoagulants considering antithrombotic activity of this species (Rose *et al.*, 1990; Tyler, 1994). Although, adverse effects due to *Gingko biloba* extract are rare but can include gastrointestinal disturbances, headache, and allergic skin rash (Tyler, 1994; Z'Brun, 1995). *A. hippocastanum* may cause gastrointestinal irritation. Parenteral administration of aescin could induce anaphylactic reactions, hepatic and renal toxic effects (Hellberg, Ruschewski, and Vivie, 1975; Takegoshi *et al.*, 1986; Tyler, 1994; Voigt *et al.*, 1978). *A. hippocastanum* extract is known as one of the components of venocuran, a drug marketed for venous disorders treatment (now out of market). In 1975, venocuran was discovered to cause a pseudolupus syndrome characterized by recurrent fever, myalgia, arthralgia, pleuritis, pulmonary infiltrates, pericarditis, myocarditis, and mitochondrial antibodies in the absence of nuclear antibodies after prolonged treatment (Grob *et al.*, 1975; Wälli, Grob, and Müller-Schoop, 1981).

## 8.5 Preclinical Models for Cardiovascular Investigation

Animal models provide preclinical assessment for the study of medicinal plants with cardiovascular activities. The potential effect of a specific plant on the cardiovascular system can be evaluated by obtaining and application of the crude extract on different models (Cechinel Filho and Yunes, 1998). Animal models that closely resembles the operation and phylogenetic of the human body are widely used in pre-clinical analysis (Fagundes and Taha, 2004). Several species of mammals have been used, among them the order of rodents, including rats (Andrade *et al.*, 2015), mice (Florentino *et al.*, 2016) and guinea pigs have been highlighted.

The advantages and limitations of each model applied depends on the similarity of pathological processes in which the animal model represents and also the different parameters that will be analyzed. The effects of an active principle extracted from a plant is analyzed in different ways on a wide range of cardiovascular parameters (Khan *et al.*, 2015; Ventrella *et al.*, 2015). The blood pressure average, heart rate, cardiac contractility and vascular alterations in blood flow are among the most commonly recorded parameters.

In particular, the use of transgenic and knockout animals have disseminated widely as an excellent tool for molecular analysis and cellular mechanisms involved in evaluating the effectiveness of new cardioactive drugs (Zadelaar *et al.*, 2007; Zaragoza *et al.*, 2011). Among the various models of cardiovascular disease, hypertension, atherothrombotic, abdominal aortic, aneurysms and heart failure are the targets of therapeutic cardioactives plants (Zaragoza *et al.*, 2011). *In vivo* verification of cardioactive actions of plants depends on the routes of administration, such as intravenous, oral, subcutaneous and intramuscular (Heikal *et al.*, 2016; Turner *et al.*, 2011). In another route, direct action of cardioactive drugs can be assessed individually and exclusively on the cardiovascular system, without interference from other systems, using techniques such as isolated organs (isolated heart and vessels) (Sá *et al.*, 2014).

Thus, different techniques and animal models are used in pre-clinical evaluation of plants with potential cardioactive. The choice of the ideal model of administration for roads and the parameters evaluated are of fundamental importance in this process. In particular, the action of cardioactive plants in the treatment and prevention of cardiovascular disease has been the subject of numerous investigations (Sá *et al.*, 2014; Andrade *et al.*, 2015; Zaragoza *et al.*, 2011). Continuing research is necessary to elucidate the pharmacological activities of the many cardiopotent herbal medicines and to stimulate future pharmaceutical development of therapeutically beneficial herbal drugs.

## Chapter 9

# Pharmacokinetics Profile of Medicinal Plants and Potential Interactions with Xenobiotic

*James O. Fajemiroye and Luiz Carlos da Cunha*

## ABSTRACT

*The general assumption of healthy living and safety with the natural product treatments has increased uncontrolled application of medicinal plants. This has led to increase in the cases of adverse reactions being reported. The high statistics of medicinal plants that fails to enter global market through safety concerns could be associated with poor understanding of their pharmacokinetic profiles. Botanical preparations or active principles can be absorbed and transported to target site of action. It is desirable that toxic metabolites of these active principles are eliminated from the body without producing adverse side effects. Pharmacokinetic data is important for understanding herbs-pharmaceuticals interactions and helps to elaborate the relationship between intensity and time course of pharmacology and the toxicological effects of phytochemicals in the human body. This chapter reports pharmacokinetic information on some medicinal plant species, interaction with conventional drugs, P-glycoprotein efflux pump and enzyme systems CYP450 which could affect their bioavailability or that of other xenobiotics. The aspect of CYP450 polymorphism is discussed as individual hepatic detoxification status, and it is a necessary prerequisite to therapeutic interventions and safe use of medicinal plants. Keywords such as medicinal plants, CYP450, glutathione, glucuronidation, sulfation, sulfate conjugation, sulfotransferase, methylation, methyltransferase, acetyla tion, n-acetyltransferase, P-glycoprotein were used to search Google Scholar, PubMed, Science Direct among others.*

## 9.1 Introduction

Pharmacodynamics (effect of drug on the body) and pharmacokinetics (effect of body on the drug) are often the main focus in classical pharmacology. The effect of medicinal plants on the body are often reported locally, however, little attention have been paid to the temporal changes in quality and quantity of botanicals as a result of its interaction with biological molecules. This may not be unconnected with the challenges that are associated with the complexity of medicinal plant chemistry, multiplicity of components, inability to identify biological markers and resulting metabolites. As a result, there is very little pharmacokinetics data for numerous medicinal plants with traditionally application (He *et al.*, 2010). The qualitative and quantitative temporal analysis of a given dose of medicinal plants preparation is influenced by pharmacokinetic parameters such as the absorption, distribution, metabolism and excretion of its various components. Pharmacokinetics knowledge of medicinal plants can provide valuable information towards prediction of potential interactions between botanicals and other xenobiotics (pharmaceuticals); and also promote safe and effective applications.

Potential interactions between botanicals and other xenobiotics (pharmaceuticals) could come from various sources, including clinical trials, preclinical trials, post licensing drug monitoring, controlled trials on healthy subjects required for drug licensing, *in vitro* or *in vivo* studies on animal or human cell lines or tissues, and Adverse Drug Reports (ADR's). The extent of drug interactions with medicinal plants can be determined by factors relating to co-administered drugs (dose, dosing regimen, administration route, pharmacokinetic and therapeutic range), medicinal plants (species, dose, dosing regimen, and administration route) and patients (genetic polymorphism, age, gender and pathological conditions) (Dresser *et al.*, 2000). The marked variation in drug interactions among individuals as a result of polymorphism or interindividual differences are often associated with the activities of drug metabolizing enzymes and transporters (Zhou, S. *et al.*, 2003).

The human drug-metabolizing enzymes such as Cytochrome P450 (CYP450) enzymes (located in the smooth endoplasmic reticulum of the liver and other extra hepatic tissues) and P-glycoprotein (P-gp) efflux pump are important biological systems which are involved in pharmacokinetic drug interactions. Several CYP450 including CYP1A2, CYP2E1 (metabolizes ethanol to acetaldehyde, acetaminophen to N-acetyl–*p*–benzoquinone(NAPQ), activates some carcinogens, procarcinogens and toxicants, CYP2E1 also has the ability to produce reactive intermediates, leading to the formation of free radicals such as superoxide, hydroxyl radical, and lipid peroxides); CYP2C9 (metabolizes nonsteroidal anti-inflammatory ibuprofen, antihypertensive losartan, antidepressant fluoxetine, antiepileptic phenytoin, anti-hypercholesterolemic fluvastatin, etc); CYP2C19 (metabolizes proton pump inhibitor omeprazole, tricyclic antidepressant amitriptyline, selective serotonin reuptake inhibitor fluoxetine, benzodiazepine diazepam and the barbiturate phenobarbital); CYP2D6 (metabolizes beta-blockers propafenone and timolol, the antidepressant amitriptyline, the antipsychotic haloperidol and risperidone, and the antihistamine chlorphenamine),and CYP3A4/5/7 (metabolizes macrolide antibiotics, antiarrythmics, benzodiazepines, immune modulators, HIV antivirals, antihistamines, calcium channel blockers and HMG CoA reductase inhibitors) have been isolated and characterized (Gibson and Skett,2001; Baxter and Stockley, 2008; Berka *et al.*, 2011; Zhou *et al.*, 2009; Anzenbacher and Anzenbacherova, 2001; Neafsey *et al.*, 2009).

The most abundant families of CYP metabolizing enzymes that account for metabolism of the majority of drugs are CYP1A2, CYP2C, and CYP3A4 isoforms (Atkinson,2012). CYP3A4 / 5 is responsible for about 50% of drug metabolism while CYP2D6 and CYP2C8 / 9 metabolize more than 20% of drugs. These enzymes are involved in redox reactions such as aromatic and aliphatic hydroxylation, epoxidation, N-dealkylation, O dealkilation, S-dealkylation, oxidative deamination, N-oxidation, S-oxidation, phosphothionate oxidation, dehalogenation, and Alcohol oxidation, hydrolysis and hydration among others (Ionescu and Caira, 2005). The evidence for interactions between these systems and medicinal plant extracts or active principles have been well established.

## 9.2 Pharmacokinetic Profile and Potential Medicinal Plants-Drug Interactions

The liver metabolic enzymes are involved in Phase 1 (functionalization reactions mediated by cytochrome P450) and Phase 2 (glutathione conjugation, glucuronidation, sulfation, methylation and acetylation) metabolism (Mazzari and Prieto, 2014). During Phase 1 reactions, polar functional groups are added to the xenobiotic drug for Phase 2 metabolism (Ionescu and Caira, 2005). The metabolic activity of these enzymes as well as the effects of some P-glycoprotein activity could be inhibited, induced or potentiated by medicinal plants. Hence, concomitant use of these plants with a conventional drug could predispose the latter to rapid or slow metabolism and subsequently reduce or increase the bioavailability of other xenobiotics thereby leading to medicinal plant-drug interactions (Hu *et al.*, 2005). Differential responses to drug among individuals have been associated with the variability of CYP content (Mazzari and Prieto, 2014). The genetic polymorphisms within CYPs that is capable of affecting the metabolism of xenobiotics could be attributed to changes in drug response and increased risk of ADRs (Zhou *et al.*, 2009).

For instance, medicinal plants that interfere with the activity of CYP1A2 (Table 9.1) could influence the metabolism of drugs such as acetaminophen (analgesic), propranolol (antihypertensive beta- blocker), clomipramine (antidepressant), Warfarin (anticoagulant) etc which are substrate of this enzyme. The biotransformation of acetaminophen by CYP1A2, CYP3A4, and CYP2E1 to toxic compound NAPQ needs to be followed by detoxification by conjugation with glutathione. An overdose of acetaminophen could lead to the saturation of this metabolic route. Accumulated NAPQ could bind to other biomolecules, resulting in hepatic cell damages (Lee *et al.*, 2001). Concomitant ingestion of acetaminophen with *Allium sativum* and *Curcuma longa* could promote accumulation of NAPQ and increase the toxicity of this drug as a result of CYP1A2 induction. In contrary, consumption of medicinal plants such as *Phyllantus amarus*, *Mormodicacharantia*, *Eucalyptus globulus*, *Glycine max*, *Harpagophytum procumbens*, *Mentha piperita*,*Trifolium pratense* and *Punica granatum* could lead to inhibition of CYP1A2 enzyme and decrease levels of this toxic metabolite (Mazzari and Prieto, 2014). Grapefruit (*Citrus × paradisi*) juice interaction with medications such as dihydropyridine, felodipine (Lundahl et al., 1997 Bailey *et al.*, 1992, 1993, 1996, 1995, 1991 Lown *et al.*, 1997 Lundahl *et al.*, 1995, 1997 Edgar *et al.*, 1992), nisoldipine, nimodipine (Fuhr

**Table 9.1: List of Medicinal Plants with Reported Effects on different Enzymes Systems (Cytochrome P family (CYP), glutathione (GLU), UDP-glucuronosyl-transferases (UGT), P-glycoprotein (GLP)**

| Plant Species/Family | CYP1A2 | CYP2C | CYP2D6 | CYP2E1 | CYP3A | UGT | GLU | GLP |
|---|---|---|---|---|---|---|---|---|
| Achille amillefolium - Asteraceae | | | | | | | ↑ (Potrich et al., 2010) | ↓ (Haidara et al., 2006) |
| Allium sativum - Allaceae | ↑ (Le Bon et al., 2003) | ↑a, ↓ab (Foster et al., 2001; Ho et al., 2010) | – (Markowitz et al., 2003) | ↓ (Le Bon et al., 2003) | –, ↓ (d, e, f) [Foster et al., 2001; Hajda et al., 2010] | ↑ (Ip and Lisk, 1997) | ↑ (Ip and Lisk, 1997) | ↑ (Hajda et al., 2010) |
| Aloe vera barbadensis - Aloaceae | | | – | | | | ↑, ↓ (Kaithwas et al., 2011; Hegazy et al., 2012) | |
| Anacardium occidentale - Anacardiaceae | | | – | | | | ↑ (Singh et al., 2004) | |
| Baccharis trimera - Asteraceae | | | – | | | | ↓ (Nogueira et al., 2011) | |
| Bauhinia forficata - Caesalpiniaceae | | | – | | | | ↓ (Damasceno et al., 2004) | |
| Bauhinia variegata - Caesalpiniaceae | | | – | | | | ↑ (Rajkapoor et al., 2006) | |
| Calendula officinalis - Asteraceae | | | – | | | | ↑ (Preethi and Kuttan, 2009) | |
| Chamomilla recutita - Zingiberaceae | | | | | –d (Budzinski et al., 2000) | | ↑ (Al-Hashem, 2010) | |
| Croton cajucara - Euphorbiaceae | | | | | | | ↑ (Rabelo et al., 2010) | |
| Curcuma longa - Myrtaceae | ↑ (Thapliyal et al., 2002) | | | – (Salama et al., 2013) | –d (Graber-Maier et al., 2010) | ↓ (Naganuma et al., 2006) | ↑ (Rong et al., 2012) | – (Graber-Maier et al., 2010) |

**Table 9.1–Contd...**

| Plant Species/Family | CYP1A2 | CYP2C | CYP2D6 | CYP2E1 | CYP3A | UGT | GLU | GLP |
|---|---|---|---|---|---|---|---|---|
| Cynara scolymus - Asteraceae | | | | – | | | –, ↑ (Miccadei et al., 2008) | |
| Eucalyptus globulus | ↓ (Unger and Frank, 2004) | ↓a,b (Unger and Frank, 2004) | ↓ (Unger and Frank, 2004) | | ↓d (Unger and Frank, 2004) | | | |
| Foeniculum vulgare - Apiaceae | | | | | ↓d (Subehan et al., 2006, 2007) | | ↑ (Zhang et al., 2012) | |
| Glycine max - Leguminosae | ↓ (Shon and Nam, 2004) | | | – (Shon and Nam, 2004) | | | ↑ (Barbosa et al., 2011) | |
| Harpagophytum procumbens - Pedaliaceae | ↓, – (Unger and Frank, 2004; Modarai et al., 2011) | ↓a,b, –a (Modarai et al., 2011) | ↓, – (Modarai et al., 2011) | | –, ↓d (Unger and Frank, 2004; Modarai et al., 2011) | | | |
| Mentha pulegium - Lamiaceae | | | | | | | ↑ (Alpsoy et al., 2011) | |
| Mentha piperita - Lamiaceae | ↓ (Unger and Frank, 2004) | ↓ab (Unger and Frank, 2004) | ↓ (Unger and Frank, 2004) | | ↓d (Unger and Frank, 2004) | | ↑ (Sharma et al., 2007) | |
| Mikania glomerata - Asteraceae | | | | | | | – (Barbosa et al., 2012) | |
| Mormodica charantia - Cucurbitaceae | | | | | ↓d (Raza et al., 1996) | | ↑ (Raza et al., 2000, 1996) | |
| Phyllanthus amarus - Euphorciaceae | ↓ (Hari Kumar and Kuttan, 2006) | | ↓ (Hari Kumar and Kuttan, 2006) | | ↓d, e, f (Hari Kumar and Kuttan, 2006) | | ↑ (Bhattacharjee and Sil, 2006; Manjrekar et al., 2008) | |

Contd...

**Table 9.1–Contd...**

| Plant Species/Family | CYP1A2 | CYP2C | CYP2D6 | CYP2E1 | CYP3A | UGT | GLU | GLP |
|---|---|---|---|---|---|---|---|---|
| Psidium guajava - Myrtaceae | | | | | | | ↑ (Tandon et al., 2012) | |
| Punica granatum - Lythraceae | ↓ (Faria et al., 2007a) | ↑ (Hanley et al., 2012) | ↓(Usia et al., 2006) | | ↓d, e, f (Faria et al., 2007a) | | ↑, ↓ (Faria et al., 2007b; Dassprakash et al., 2012) | |
| Ruta graveolens - Rutaceae | | | | | | | ↑ (Ratheesh et al., 2011) | |
| Trifolium pretense - Fabaceae | ↓ (Unger and Frank, 2004) | ↓a,b (Unger and Frank, 2004) | ↓(Unger and Frank, 2004) | | ↓d (Budzinski et al., 2000) | | | |
| Uncaria tomentosa - Rubiaceae | | | | | ↓d (Budzinski et al., 2000) | | | |
| Zingiber officinale - Zingiberaceae | | | ↓ (Kimura et al., 2010) | | | | | |

**Symbols-abbreviations**: ↑, Enzyme induction; ↓, Enzyme inhibition; -, No Effect; a, CYP2C9; b, CYP2C19; d, CYP3A4; e, CYP3A5; f, CYP3A7; GLU, glutathione; UGT, UDP-glucuronosyl-transferases; GLP, P-glycoprotein.

*et al.*, 1994), nicardipine (Uno *et al.*, 1997), nitrendipine (Bailey *et al.*, 1992 Soons *et al.*, 1991), nifedipine (Rashid *et al.*, 1993, 1995 Sigush *et al.*, 1996), and amlodipine (Josefsson *et al.*, 1996 Vincent *et al.*, 1997), that are substrates for CYP3A4 have been extensively studied.

The glutathione conjugation plays important role in the removal of toxic electrophilic compounds from the body (Ionescu and Caira, 2005) while the final conjugated product is then eliminated from the organism (Sies and Ketterer, 1988). Meanwhile, medicinal plants could deplete glutathione levels and predispose individual to the toxic effect of drugs that are consumed concomitantly with this species (see table below). Hence, the changes on the glutathione conjugation mechanism could increase chances of plant-drug interactions (Mazzari and Prieto, 2014). Conjugation of xenobiotics with chemical groups such as alcohols, phenols, hydroxylamines, carboxylic acids, amines, sulphonamides, and thiols could also be achieved by glucuronidation reaction (Gibson and Skett, 2001). Glucuronidation of some phase 1 metabolites facilitates their elimination from the organism (Kuehl *et al.*, 2005). Glucuronidation mechanism involves the formation of glucuronide through the reaction catalyzed by UDP-glucuronosyltransferases(UGTs) between the electrophilic C-1 atom of the pyranose acid ring of the co-factor UDPGA (uridine 5-diphosphate-glucuronicacid) with the substrate (Mazzari and Prieto, 2014). Hence, an increases, decrease or inhibition in the expression of UGTs in the liver by medicinal plants consumption could compromise the pharmacokinetic of class of drugs that undergoes glicuronidation prior to elimination. Medicinal plants effects on P-glycoprotein (GLP) activity could also lead medicinal plant-drug interaction with profound pharmacokinetic consequences. GLP acts as an efflux pump in transporting metabolites and xenobiotics across biological membranes out of the cells that can result in pharmacokinetic alterations (Williamson *et al.*, 2009).

## 9.3 Pharmacokinetic Profile of Phytoconstituent(s) Isolated from Medicinal Plants

The therapeutic benefits from medicinal plants can be attributed to the presence of numerous active principle. Hence, pharmacokinetic profile of the plant extracts and their isolates are important to the monitoring of their pharmacological actions. For instance, a single glass of grapefruit juice could increase oral bioavailability, enhance the beneficial or adverse effects of drugs. Grapefruit juice is known to inhibit presystemic drug metabolism mediated by CYP3A isoforms. The inhibition could lead to appreciable changes in therapeutic index of other drugs that are metabolized by this enzyme. The active ingredient(s) in this juice or any other herbal preparations are largely responsible for the interaction with other drugs or xenobiotics. For instance, flavonoid like Naringin in grapefruit juice, can inhibit drug oxidative metabolism (Buening *et al.*, 1981), thereby producing interaction with other drugs (Yee *et al.*, 1995). The *in vitro* metabolism of felodipine and nifedipine has been reported to be inhibited by Naringin (Deslypere *et al.*, 1991). Pharmacokinetic profile could be considered as a blueprint for the medicinal plant. Both qualitative and quantitative analysis could corroborate therapeutic and toxicological effects of medicinal plants. Few among the secondary metabolites with the description of their pharmacokinetic parameters are listed below and on Table 9.2.

## Visnagin

Visnagin, a furanocoumarins derivative and isolate of *Ammi visnaga* L., is a calcium channel blocker with negative chronotropic and inotropic effects and reduces peripheral vascular resistance (Mehta *et al.*, 2015). The application of *Ammi visnaga* in the treatment of urolithiasis (kidney stone formation) may be associated with the activity of this compound. Chromatography conditions for quantitative determination of visnagin in a rat plasma (after oral administration) have been reported. Pharmacokinetic parameters after oral administration of visnagin are provided in Table 9.2 (Mehta *et al.*, 2015)

## 3-n-Butylphthalide

3-n-Butylphthalide that seems to be a promising new drug for the treatment of ischemic cerebral diseases has been detected in *Apium graveolens*, *Ligusticum sinensis*, and *Ligusticum wallichii* (Mehta *et al.*, 2015). The chromatography conditions for the determination of 3-n-butylphthalide in rat plasma and pharmacokinetic parameters after intravenous administration of 3-n-butylphthalide (5 mg/kg) have been reported as shown in Table 9.2.

## Atractylenolide I

Atractylenolide I isolated from *Atractylodes macrocephala* Koidz showed anticancer effect. HPLC-MS/MS method was reported for quantification of atractylenolide I in ethanolic extract of *A. macrocephala* and pharmacokinetic parameters have been reported as shown in Table 9.2 (Mehta *et al.*, 2015).

## Artemisinin

Artemisinin is an antimalarial agent and secondary metabolite isolated from *Artemisia annua* L. This species has been used for the treatment of fever and malaria. The pharmacokinetic parameters of artemisinin are provided in Table 9.2 (Mehta *et al.*, 2015).

## Aristolochic acids

Aristolochic acids (AAs) has been detected in *Aristolochia fangchi*. This compound has anti-inflammatory, analgesic, antitussive, and antiplatelet aggregation. The pharmacokinetic parameters of this compound has been determined using Ultra high-performance liquid chromatography (Mehta *et al.*, 2015).

## Mangiferin

Mangiferin which possesses antioxidant, antiviral, and anticancer activitities has been isolated from the root of *Anemarrhena asphodeloides* Bung. The pharmacokinetic parameters of Mangiferin has been reported as shown in Table 9.2 (Mehta *et al.*, 2015).

**Table 9.2: Pharmacokinetic Parameters of some Medicinal Plant Isolates**

| Isolates | Cmax | Tmax | $t_{1/2}$ | AUC | CL/F | Ke | Vd |
|---|---|---|---|---|---|---|---|
| Visnagin | 2969 ng/mL | 0.33 h | | 11.9 mg h/L | 0.84 L/h kg | | |
| 3-n-Butylphthalide | | | 2.62 h | 1140.16 ng h/mL | 3.67 L/h kg | | 1.22 L/kg |
| Atractylenolide I | 7.99µg/L | 0.81 h | 1.94 h | 22.2 µg h/L | | 0.365 | 2768.6 L/kg |
| Artemisinin | 240 ng/mL | 0.6 h | 0.9 h | 336 ng h/mL | | | |
| Aristolochic acids | 7249.3 ng/mL | 30 min | 234.6 min | 716,9 ng mL/min | 2.9 mL/min kg | | |
| Mangiferin | | | 28.5 min | 122.9 ng mL/min | 90.2 mL/min kg | | |
| Indirubin | 201 ng/mL | 0.017 h | 1.0 h | 308 ng mL/min | | 0.670 | |
| Cudratricusxanthone B | | | | | | | |
| Nobiletin | 1.78 µg/mL | 1.00 h | 1.80 h | 7.49 µg. h/mL | | | |
| Catechin | | | | 109.7 ng h/mL | | | |
| Epicatechin | | | | 67.66 ng h/mL | | | |
| Curcumin | 0.36 µg/mL | | | 7.2 ng mL/min | | | |
| Protodioscin | 70 µg/mL | | 78 min | 732 µg mL/min | 0.64 mL/min kg | 0.0089 | |
| Isorhamnetin | 195.96 ng/mL | 7.21 h | | 1153.66 ng h/mL | | 0.0321 | |
| Quercetin | 179.21 ng/mL | 1.21 h | | 1368.26 ng h/mL | | 0.0541 | |
| kaempferol | 180.23 ng/mL | 6.32 h | | 1139.59 ng h/mL | | 0.1641 | |
| Lasiodonin | 1300.717 ng/mL | 105 min | 40.9 min | 96990.82 ng mL/min | | | |
| Oridonin | 1916.333 ng/mL | 105 min | 45.1 min | 142768 ng mL/min | | | |
| Ponicidin | 1582.383 ng/mL | 120 min | 40.3 min | 115004 ng mL/min | | | |
| Rabdoternin A | 385.011 ng/mL | 105 min | 58.04 min | 28958 ng mL/min | | | |
| Indolinone | | 4.30 min | | 561 ng h/mL | 3.38 L/h kg | 9.53 | |
| Mangiferin | 301.3 µg/mL | 2.5 h | 3.2 h | 1855.0 µg L/min | | | |

*Contd...*

**Table 9.2–Contd...**

| Isolates | Cmax | Tmax | $t_{1/2}$ | AUC | CL/F | Ke | Vc |
|---|---|---|---|---|---|---|---|
| **Kakkalide** | 0.26 µg/mL | 0.25 h | 0.95 h | 0.24 µg mL/min | 1021.7 L/h kg | | |
| **Paeonol** | 0.73 µg/mL | 9.00 min | | 40.66 µg mL/min | 1016.8 L/h kg | | 47,170 mL/kg |
| **Tetrahydropalmatine** | 435.8 µg/mL | 1.50 h | 6.68 h | 3450.1 ng h/mL | | | |
| **Protopine** | 347.9 µg/mL | 3.50 h | 4.98 h | 2987.0 ng h/mL | | | |
| **Palmatine** | 8.53 µg/mL | 1.92 h | 12.84 h | 57.85 ng h/mL | | | |
| **Rhein** | 42.66 mg/mL | 0.9h | | 80.28 mg h/L | 0.78 L/h kg | | |
| **Oridonin** | | | | 7.96 µg mL/min | 1.56 L/h kg | | 1.83 L/kg |
| **Chamaechromone** | 795.9 ng/mL | 11.3 h | 30.0 h | 6976.7 ng h/mL | 13,731.8 L/h kg | | |
| **Tetrandrine** | 237.1 ng/mL | 6.0 h | 20.6 h | 6279.2 µg mL/min | | 0.034 | |
| **Solamargine** | | | 3.5 h | 242.41 ng h/mL | 3.81 L/h kg | | 20.16 L/kg |
| **6-Gingerol** | 0.933 µg/mL | 1.167 h | 3.6 h | 1.689 µg mL/h | 57.43 L/h kg | | |
| **8-Gingerol** | 0.092 µg/mL | 0.833 h | 1.1 h | 0.177 µg mL/h | 60.59 L/h kg | | |
| **10-Gingerol** | 0.156 µg/mL | 0.361 h | 1.6 h | 0.222 µg mL/h | 154.898 L/h kg | | |
| **6-Shogaol** | 0.111 µg/mL | 1 h | 1.1 h | 0.14 µg mL/h | 133.652 L/h kg | | |

Mehta *et al.*, 2015; Vanachayangkul *et al.*, 2009; Niu *et al.*, 2008; Diao *et al.*, 2013; Li *et al.*, 2012; Rath *et al.*, 2004; Kuo *et al.*, 2010; Lai *et al.*, 2003; Deng *et al.*, 2008; Pi *et al.*, 2010; Singh *et al.*, 2011; Wang *et al.*, 2007; Chen *et al.*, 2010.

**Abbreviations:** Cmax, Peak plasma concentration; Tmax, Time of peak plasma concentration; AUC, Area under the concentration-time curve) (trapezoidal rule):-, Apparent clearance; Vd, Apparent volume of distribution; $t_{1/2}$; Elimination half-life; Ke, Elimination rate constant; ; CL/F, Apparent clearance.

## Indirubin

Indirubin which has been detected in *Baphicacanthus cusia* (Nees) Bremek has antileukemic, antiproliferative and anti-inflammatory activity. The pharmacokinetic parameters of indirubin has been detected using HPLC and reported as shown in Table 9.2 (Mehta *et al.*, 2015).

## Nobiletin

Nobiletin, a polymethoxylated flavone, is commonly found in citrus fruit peels such as *Citrus depressa* (shiikuwasa), *Citrus sinensis* (oranges), and *Citrus limon* (lemons). Anti-inflammatory, antitumor proliferation, antitumor invasion, and neuroprotective properties of Nobiletin have been reported. The pharmacokinetic parameters of this compound are shown in Table 9.2 (Mehta *et al.*, 2015).

## Cudratricusxanthone B

Cudratricusxanthone B has been isolated from *Cudrania tricuspidata* (Carr.) Bur., a species being used for gastric carcinoma. A HPLC-ESI-tandem MS has been used to evaluate the pharmacokinetic parameters of this compound as shown in Table 9.2 (Mehta *et al.*, 2015)

## Catechins

The catechins are often used to treat atherosclerosis and cancer. Both catechin and epicatechin have been isolated from *Cynomorium songaricum*. The LC-MS/MS method has been used for simultaneous determination of pharmacokinetic parameters of catechin and epicatechin as shown in Table 9.2 (Mehta *et al.*, 2015).

## Curcumin

Curcumin, a phenolic substance derived from *Curcuma longa* L., has been used as a natural food colouring agent, curry powder, anticancer, antiviral, anti-infectious and anti-amyloidogenic agents. The pharmacokinetics parameters have been determined by using LC-MS/MS with an orthogonal Z-spray electrospray interface system (Mehta *et al.*, 2015).

## Protodioscin

Protodioscin is a typical example of a furostanol saponin, which is isolated from the roots of *Dioscorea nipponica* Makino. Protodioscin is a potent anticancer agent. the pharmacokinetics parameters have been reported on Table 9.2 (Mehta *et al.*, 2015).

## Quercetin, Kaempferol, and Isorhamnetin

Quercetin, kaempferol, and isorhamnetin are isolates of *Ginkgo biloba* are responsible for the free radical scavenging effects of *G. biloba*. The LC-20AB with SPDM20A (Shimadzu) has been used to determine the pharmacokinetic parameters of these compounds as shown on Table 9.2 (Mehta *et al.*, 2015).

## Lasiodonin, Oridonin, Ponicidin, and Rabdoternin A

Lasiodonin, oridonin, ponicidin, and rabdoternin A are active principles that have been isolated from *Isodon rubescens* (Hemsl.), a species that is used in folk medicine for respiratory, gastrointestinal inflammatory and cancer diseases. HPLC-ESI-MS method with positive ionization mode has been used to determine pharmacokinetic profiles of these compounds as shown on Table 9.2 (Mehta *et al.*, 2015).

## Indolinone

Indolinone, an alkaloid present in the dried roots of *I. indigotica* L. Indolinone is an inhibitor of mast cell degranulation and blocker of immunoglobulin E (IgE) mediated degranulation of sensitized mast cells. An UPLC-MS/MS method with positive ionization mode has been used to determine the pharmacokinetic parameters of this compound as shown on Table 9.2 (Mehta *et al.*, 2015).

## Mangiferin

Mangiferin has been isolated from *Mangifera indica*. This compound has shown promising anti-diabetic activity. An ACQUITYTM UPLC/MS system (Waters Corp., Milford, MA, USA) has been used to determine its pharmacokinetic parameters as shown on Table 9.2 (Mehta *et al.*, 2015).

## Kakkalide

Kakkalide, an isoflavone found in extracts from the dried flower of *Pueraria lobata* (Willd.). *P. lobata* has been used to treat symptoms that are associated with excessive alcohol intake, such as drunkenness, headache, red face, and liver injury (Mehta *et al.*, 2015).

## Paeonol

Paeonol is an active principle found in the root cortex of *Paeonia suffruticosa* A. Preparation from this species is often prescribed for the treatment of pain and inflammatory ailments. A HPLC-DAD method has been used to determine pharmacokinetic profile of paeonol (Mehta *et al.*, 2015).

## Tetrahydropalmatine, Protopine, and Palmatine

Tetrahydropalmatine, protopine, and palmatine have been isolated from *Corydalis* decumbentis, a species commonly used for the treatment of hemiplegia, rheumatoid arthritis, infantile residual paralysis, dementia, hepatotoxicity and pain. A LC-ESI-MS with a positive ion mode has been used to determine the pharmacokinetic profile of these compounds as shown on (Table 9.2: Mehta *et al.*, 2015).

## Rhein

Rhein, an active principle found in plants such as Aloe spp. It has antitumor, anti-inflammatory, antibacterium, and renal protection properties. A HPLC with

Shimadzu RF-10A fluorescence detector was used to determine pharmacokinetic parameters of this compound (Table 9.2: Mehta *et al.*, 2015).

## Oridonin

Oridonin is a diterpenoid compound isolated from the *Rabdosia rubescens*. This species has anti-inflammation, antibacterial, and antitumor effects. Reverse phase-HPLC has been used to determine pharmacokinetic parameters of oridonin (Table 9.2: Mehta *et al.*, 2015).

## Chamaechromone

Chamaechromone, a biflavone found in dried roots of *Stellera chamaejasme* L., has anti-inflammatory, antiviral, cytotoxic, and antioxidant activity. A LC-MS with a positive ESI in MRM mode was used to determine the pharmacokinetics parameters of this compound (Table 9.2: Mehta *et al.*, 2015).

## Tetrandrine

Tetrandrine, a bisbenzyl isoquinoline alkaloid, has been isolated from the root of *Stephania tetrandra*. It has anti-inflammatory, antiallergic, antioxidant, and antifibrogenetic activities. The LC/MS/MS method has been used to determine the pharmacokinetic parameters of tetrandrine (Table 9.2: Mehta *et al.*, 2015).

## Solamargine

Solamargine is a steroid alkaloid glycosides found in *Solanum spp.* are well known for its antitumor activity. The HPLC method coupled with Shimadzu LCMS-2010A quadrupole MS by an ESI interface has been used to determine the pharmacokinetic parameters of solamargine (Table 9.2: Mehta *et al.*, 2015).

## Oleoresin

Oleoresin is the nonvolatile pungent component from the dried rhizome of *Zingiber officinale* Roscoe (Ginger), a popular spice and flavoring agent and dietary supplement for nausea and motion sickness. The major constituents of oleoresin include 6-gingerol, 8-gingerol, 10-gingerol, and 6-shogaol. A HPLC-MS method with positive ionization interface was used to determine the pharmacokinetic parameters of ginger oleoresin (Table 9.2: Mehta *et al.*, 2015).

In summary, it is important to consider the maximum possible variables to obtain good results in pharmacokinetics. We highlight some aspects related to animal testing, such as age and weight of the animals (in fasting condition); and routes of administration model [at least oral (crude extracts or pure compounds) and intravascular (pure compounds) routes; dosing schedule (single or multiple dose(s)]. The time and route of collection of blood samples; dose and solubility of the extract or isolated compounds; and possibility of interference/interaction with other constituents are also very relevant in animal testing. A very important aspect is that we must always distrust the solubility and / or permeability of the active ingredient in the study, because many plant constituents are in conjugated form and may exhibit low oral bioavailability. Hence, there is need for an intravascular route of administration.

# Chapter 10

# Scientific Reporting

*James O. Fajemiroye and Christianah A. Elusiyan*

## ABSTRACT

*Scientific report remains one of the means to demonstrate understanding of the science being studied and practiced. Scientists are often faced with the need to present their findings for evaluation, funding, promotion, scientific meeting or publication in professional scientific reports. In all these, the work of scientists or researchers are often exposed to criticisms, praises, acceptance or rejection by interest groups or concerned authorities. Scientists often spend quality time writing proposals, planning research, undertaking experiments, analyzing data, tracing research, and reading related articles but at times fail to pay attention to effective communication of their findings with good report. The scientific report should be considered as an important historical document that provides evidence of scientific effort and dedication of researcher. Scientific writing should be clear, unambiguous and accurate, contain all necessary details, concise, conform to style of reporting with sound use of technical terms and grammar, consistent spelling, structure, font, format, decimals, abbreviations, layout among others.*

## 10.1 General Introduction

In addition to the experimental design and conduction, reporting of experimental data is a very important phase in basic researches. Reporting of experimental data could be a tricky thing to do based on different practices and conventions that have been developed over time. Hence, what is regarded as a suitable format of writing report may vary from author to author. Generally, researchers are often expected to write reports and to present findings in written, verbal or audio-visual forms. This chapter will focus more on report writing. The writing process is divided into three phases: pre-writing phase- (planning), writing phase and post-writing phase (editing). Good report remains a means by which researcher or author can connect to their reader. If the overall point or 'message' of the report is unclear, the report is badly structured, without logical and sequential progression of ideas, too long, boring and does not engage the reader, inappropriate words, spelling mistakes and grammatical errors among others, there are possibility that the author may fail to connect with potential reader. Research findings need to be communicated and disseminated for them to be useful to society, and therefore scientists have an obligation to prepare understandable reports that can be later published. There are growing concerns in the medicinal plant research regarding the replicability of published research (Pashler and Wagenmakers, 2012; Ioannidis, 2005). In order to improve the quality, clarity, completeness and transparency in medicinal plant research reporting, authors of reports need to be consistent in their research approach by following a well-defined and relevant guidelines.

For replication to be possible, accurate and complete reporting of scientific methods is essential. Oversights and omissions are not uncommon in scientific reporting, and noticing them requires considerable efforts. An unreported critical detail could render experimental results impossible to interpret for the author and reader thereby making the study impossible to replicate. For instance, in *in vivo* experiments, the ARRIVE guidelines have essentially highlighted some useful checklist for good reporting (Kilkenny *et al.*, 2010). An accurate and concise description of a typical experimental report could include the title, name (s) of report author (s), abstract, graphical abstract, keywords, summary of the background, relevance of study, research objectives, methods, details of the species or strain of animal used, collection and identification of plant material, date and place of collection, voucher specimen, principal findings, appropriate schemes, figures or table, discussion and conclusions. The report should be checked for spelling and grammar to avoid any form of ambiguity or misinterpretation.

## 10.2 Contents of a Scientific Report

### Title

Writing a good title can be demanding. In some cases, good report have failed to attract all the needed approval because of poor title. It is essentially important for the author of report to spend time on the titles of their reports. The title of any scientific report could summarize the study in a concise and informative manner, facilitate retrieval of report and its uniqueness among others. The uniqueness of a report could be showcased with the name of the species or breed involved. A captivating title often makes the entire report a must read. The title is the first chance to make a good impression on potential readers. The fact that hundreds of reports are been sent out for evaluation makes the need to shorter titles. Redundancy, "filler" words, general or broad title such as "central nervous system; effect of new plant isolate" should be avoided. Author of reports should also avoid abbreviations and formulae where possible. This will generate several questions that could eventually make reading of such report unattractive. The readers are often interested in a title that is specific and gives ample idea of a specific study. According to the 19th century English writer and eccentric, Charles Caleb Colton, "writer does the most who gives his reader the most information and takes from him the least time." Nothing works better than a well-written title to make sure that the wrong reader does not waste time on the wrong report, and that the right reader does not mistakenly skip over the right report. There are different ways of choosing a title. The title could be descriptive in a phrase, complete sentence or question format. Author should bear in mind that title always creates expectation among readers or evaluator of report, avoid the use of terms that are not reflective of the objective and main findings. There should be link between the title and the body of the report. A good strategy for a good title could involve proposal of different titles and seeking of help from experienced or senior colleagues to decide which of the title fits your data most closely. A successful selection of a great title is a step ahead in sound reporting. The title is the first thing a reader sees, and so should be the last thing an author writes. Although it seems unconventional to write the title last, however, working title can be adopted prior to the completion of the report. Working title should be revised to fully reflect the content of the entire report.

## Abstract

The abstract remains an integral part of a good report. Writing an abstract requires a lot of attention. A concise and factual abstract is required. A well thought off abstract is likely to encourage a potential reader to venture further. It is important to capture the attention of the reader in the first sentence or two. The lead paragraph could begin with a sentence containing the main point of the piece followed by second most important point, etc. At the end of the first paragraph, the classic questions "who, what, where, when and why" must have all been answered. The abstract should state briefly the background or motivating factor, objective of the research, methods, materials, the principal results, conclusions and implications. An abstract is often presented separately from the article, hence, it could be considered as a stand-alone summary of report. Reference citations are often avoided within abstract. A typical abstract are limited by number of words, hence, every word must be chosen carefully. In addition, non-standard or uncommon abbreviations should be avoided, but if essential, they must be defined at their first mention in the abstract itself. Abstract can be structured. The structured abstract formalizes topical areas mentioned earlier by adding subheadings and subsections (the "structure") such as background, aim, approach, results, conclusion into the abstract. Each subsection could have one to two sentences that attempt to answer some research questions. For instance, Background: What are the issues that led to this work? Aim: What did you plan to achieve in this work? Approach: What are the experimental procedures or methods you set to achieve your aims? Results: What were the main findings of the study? Conclusions: What were your main conclusions? Why are the results important? Where will they lead? The structured abstract encourages the author to include important information, facilitate crosschecking, reading, searching and comprehension. In order to take advantage of the benefits associated with structured abstract, even if the final abstract is not going to be in the form of structured abstract, the structure abstract approach can be used first and simply delete the subheadings and combine all the lines into one paragraph. Finally, reread this new abstract, and change sentence beginnings to increase readability and flow. Another strategy to draw more attention to any scientific report could involve the use of graphical abstract. The graphical abstract could concisely summarize the contents of the report through readable and relevant image.

## Keywords

Keywords seem to be the shortest session of any report and probably either taking for granted or almost ignored. This is a dangerous thing to do as keywords are essential for any successful report. Without keywords, it may be difficult to find report and prolong its usefulness through continuous access over years. As it were, most search engines, websites or databases use the keywords found to display reports. In order words, a report without keywords may not be found by using these technological facilities. The keywords which could be up to 6 are important for indexing purposes. The selection of appropriate words out of hundreds of such within the report could be daunting. General, plural terms and multiple concepts should be avoided. It is advisable to do away with the use of abbreviations as keywords. In order to identify the kind of words that could be eligible as keywords, when such words are typed into a search engine your report should come up. The words should be unique to your report in order to reduce the chances of coming up with too many hits that are off the scope of the report. The most important keywords should come up in the title, abstract and body of the manuscript several times.

## Abbreviations

The familiarization of author with self-coined-abbreviation often makes them to assume that every reader understands what such abbreviations stand for. Technical fields are full of abbreviations whose meanings experts take for granted. In a report, it may be especially tempting to abbreviate terms to meet word count targets and to make otherwise long sentences more readable. An abbreviation (from Latin, brevis meaning short) is a shortened form of a word or phrase. It consists of a group of letters taken from the word or phrase. An abbreviation is a shortening by any method; a contraction is a reduction of size by the drawing together of the parts. A contraction is an abbreviation, but an abbreviation is not necessarily a contraction. Abbreviations have a long history, created so that spelling out a whole word could be avoided. This might be done to save time and space, and also to provide secrecy. Abbreviations are sometimes useful for long and technical terms in scientific writing; communication is usually garbled rather than clarified if, for example, an abbreviation is unfamiliar to the reader. Abbreviation overuse can reduce readability, forcing a non-specialist reader to pause and refer back to the original definition. Hence, the use of abbreviations in a report could depend on the intended readers. A pervasive abbreviation may not need to be defined as most readers will already be aware of their meaning. However, no matter how long a word could be, if it appears once in a report, there may be no need for having it abbreviated. Although a perversive abbreviation may be probably well understood by many readers, these abbreviations should still be explained when first used. Abbreviations that are unavoidable in the abstract must be defined at their first mention. Author of report should be consistent with the use of abbreviations throughout the report.

## Units

In addition to abbreviations, units are sometimes used along with technical terms and values. Author should refer to internationally accepted rules and conventions. The use of the international system of units should be encouraged. The international system of units, prefixes, and symbols could be used for all physical quantities without risk of confusion or ambiguity. These units provide a better representation of the phenomena (intensity, weight, luminosity, etc) concerned. SI units are used to a varying extent in scientific reports. The units of the *centimeter - gram - second* system remain the most common SI units. In some cases, authors could use *per* to indicate division. Per can be substituted by using either a negative index or a solidus (oblique stroke or slash). For instance, the SI unit of velocity which is metre per second could be written as $ms^{-1}$ or m/s. Author should be consistent with the space between the units and numerical values. In addition to space, the superscript, subscript, full stop and coma should be used appropriately to avoid any ambiguity.

## Introduction

An introduction is the first paragraph of a written report about your project. The introduction defines the subject of the report. It must outline the scientific purpose(s) or objective(s) for the research performed and give the reader sufficient background to understand the rest of the report. Without an introduction, it is sometimes very difficult for readers to figure out what are in the report. There needs to be a thread of an idea that they will follow. Care should be taken to limit the background to whatever is pertinent to the experiment. A sufficient scientic background could include relevant references to previous work in order to understand the motivation and context for the study, and explain the experimental approach and rationale; an explanation of how and why the animal species and model being used can address the scientific objectives and, where appropriate, the study's relevance to new discovery or advancement of knowledge. The author can review the literature, show the historical development of an idea and include the confirmations, conflicts, and gaps in existing knowledge under introduction. The specific hypothesis and experimental design pertinent to investigating the topic should be described. An hypothesis is central to the scientific method of research. It is considered as an assumption or reasonable "guess" based on what is currently known. The hypothesis is testable and can predict or propose a relationship between two or more variables. With the inclusion of various components, it is recommendable to write introduction at the end of the project to make sure that every aspect of the report is effectively captured. Any modification to the project at the development stage should be accurately reflected in the introduction. Every sentence in the introduction should be consistent with the modification. Author should avoid making introduction into a mini-review. Although there are hundreds of literature out there, author should be able to pick out the things that are most relevant and explain the rationale behind their choice. The use of appropriate literature will make the reader appreciate how much the author understand their area of research. Many people start with a broad statement and then narrow the subject matter down gradually to their specific area of interest. This is not necessarily wrong, but why bother discussing things that are not really that relevant? (Richard, 2013)

## Material and Methods

The difficulty in writing the materials and methods section of the experiments has to do with the need to provide enough detail for the reader to understand the experiment without overwhelming him or her. This section should provide answers to; What materials were used? How were they used? Where and when was the work done? The drugs and reagents that are used in the course of the research should be listed along with the name of suppliers. The solvents in which the drugs are suspended or dissolved, the precautions that are associated with the use or handling of drugs and reagents should be included. The reference drugs, test compounds and vehicle for the control group should be defined. The treatment regimen should be clearly stated. The scientific and popular name of the plants material under study should be included. Informations on the identification, collection procedure, period of collection, distribution and geographical area of collection and the voucher specimen that was deposited in the herbarium should be included. The storage, extraction procedures and phytochemical analysis should be highlighted. Under methods, researcher should provide informations on the contents of the ethical review permissions, relevant licences and national or institutional guidelines for the care and use of animals, that cover the research, brief details of the study design including the number of experimental and control groups, any steps taken to minimize the effects of subjective bias when allocating animals to treatment (*e.g.* randomization procedure) and when assessing results (*e.g.* if done, describe who was blinded and when), the experimental unit (*e.g.* a single animal, group or cage of animals). On the experimental animals, author should provide details of the animals used, including species, strain, sex, developmental stage (*e.g.* mean or median age plus age range); and weight (*e.g.* mean or median weight plus weight range), the source of animals, international strain nomenclature, genetic modification status (*e.g.* knock-out or transgenic), genotype, health/immune status, drug or test naïve, previous procedures, etc; housing type, type of cage or housing, bedding material, number of cage companions, housing conditions (*e.g.* breeding programme, light/dark cycle, temperature, type of food, access to food and water, environmental enrichment), welfare-related assessments and interventions that were carried out prior to, during, or after the experiment, sample size, the total number of animals used in each experiment, and the number of animals in each experimental group; explain how the number of animals was arrived at, provide details of any sample size calculation used, indicate the number of independent replications of each experiment and give full details of how animals were allocated to experimental groups. Where applicable (for instance in studies with genetically modified animals) the generation should also be given, as well as the details of the wild-type control group (for instance littermate, back cross etc). Relevant characteristics and health status of animals (*e.g.* weight, microbiological status, and drug or test naïve) should be reported. The details of drug formulation and dose, site and route of administration, anaesthesia and analgesia used [including monitoring], surgical procedure, method of euthanasia), details of any specialist equipment used, including supplier(s) should be reported. Information on the duration of experiment and time of day, environmental conditions (relative humidity, temperature, light intensisty, etc), home cage, laboratory, rationale for choice of specific anaesthetic, route of administration and drug dose used should be provided. Describe any modifications to the experimental protocols and the rationale behind such modifications. When procedures from previously published report are followed exactly, simply cite the work, noting that details can be found in that particular source. However, it is still necessary to describe special pieces of equipment and the general theory of the assays used. All the *in vivo, ex vivo* or *in vitro* models used should be described effectively. The inclusion and exclusion criteria of animals in tests should be clearly described. The criteria under which exclusion occurs should be determined in advance prior to allocation to experimental groups and pretest session. The number of animals excluded should be reported. Numbers below ten should not be written in numerals when they are not associated with measurements.

## Ethical Considerations

Biological experiments involving animals or humans should apply for due permission or ethical clearance for the conduction of such experiments from relevant authorities such as ethics committees. All in vivo experiments must be conducted following good laboratory practices and guidelines on the use of animals. Research work involving human studies should also be conducted according to internationally approved procedures. Verified letter of approval from such Ethics Committee or the Institutional Review Board must have been obtained before embarking on such research.

### Statistical Analysis of Experimental Data

In order to make valid inference from experimental results, data should be subjected to appropriate analysis. There are different ways to report statistics in the text, however, there are some important elements that are often included. The number of animals in each group and reasons for data exclusion should be indicated. The kind of statistical analysis (Student t-tests, ANOVA (1-way or 2-way), MANOVA, etc.) and post hoc tests (Tukey, Bonferroni, Newman keuls, etc) with a measure of precision (*e.g.* standard error or *condence* interval), title (ulcerated area, acidity, latency, etc) and unit of analysis for each dataset should be reported. Values and appropriate units can be reported within the text. Author should avoid reporting data which are already represented in a table or figure. Results in textual form should include corresponding table or figure number, statisctical analysis and level of significance. The table or figure itself should appear as soon as possible after it has been mentioned in the text. If there are a large number of significant differences to report, you can make a general statement about the level of significance at the beginning of the Results section. When reporting Statistics in the text, always present the mean, number of observations, F values from ANOVA, the probability level and the measure of variability of observations (range, standard deviation, standard error of the mean, etc.) and interactions among other factors. The number of decimal places should be consistent when presenting values of mean, standard deviation and/or the standard error of the mean of a particular parameter under analysis (e.g. arterial blood pressure, height, weight, etc). Percentages are often inserted without decimal places. In order to report Student t-tests within the text, the degrees of freedom are placed in parentheses and the significance level *e.g.* t (34) = 1.03, p <0.05. In order to report main effect using ANOVA, two degrees of freedom should be reported (degrees of freedom between-groups and degrees of freedom within-groups, separated by a comma), followed by F value and the significance level, separated by a comma (*e.g.* F(2,125) = 1.43, p < 0.01).

## Results

The result section focuses on the actualisation or otherwise of the stated objectives of the study. In short communications or reports of preliminary findings of a major study, both results and discussion sections can be merged into one section. In other words, the results are presented and discussed together.

For a more expansive experimental research or some journal format and style of presentation however, the result section stands alone. It usually contains the findings of the study presented in the form of tables, figures illustrations (such as diagrams and photographs). The results section can be further sub-sectioned into parts in order to integrate and present the several different aspects of the study clearly. Each sub-section should however be accompanied by a narrative statement or brief comments that explains the content of the tables or figures. More extensive or detailed comments on the findings of the study are reserved for the discussion section.

Figures, tables, photographs etc. must be titled and labelled clearly and completely with footnotes explaining abbreviations or symbols used.

## Discussion

The discussion should focus on the interpretation and scientific implications of results. Interpret the results, taking into account the study objectives, hypotheses, current theory and other relevant studies in the literature within a larger context that was established in the introduction. Author should comment on the study limitations including any potential sources of bias, any limitations of the animal model, and the imprecision associated with the results (Schulz *et al.*, 2010). The implications of the experimental methods or findings for the replacement, refinement or reduction of the use of animals in research should be included. Comment should be made on whether, and how, the findings of this study are likely to be translated to other species, including any relevance to human biology. Author should elaborate the relationship between the hypothesis and the results. To what degree does the data support the hypothesis. In some cases, the introduction may be revised, modified or rewritten if the results are not consistent with the initial hypothesis and overall direction of the introduction. In case the results do not support the working hypothesis, author could detail conceptual, reasoning or laboratory procedure challenges that could be associated with discrepancies or inconsistent results. The scientific basis and new hypotheis should be postulated to explain the relationship between the hypothesis and the results. Author needs to emphasize on the significance of their findings and propose prospective measure to improve similar research in the future. All the informations should be effectively organized in paragraphs with each paragraph focusing on a main idea. Author should avoid repetition of the content in introduction under discussion and the use of numbers to start sentences. All the data, detailed analysis of graphs, tables, and drawings presented in the results should be avoided under discussion.

## Conclusion, Conflict of Interests, Acknowledgements and Disclosures

Conclusion is usually a brief section where main findings are presented. The relevance of the experiment to the field of research and future project are often highlighted in this section. The source of funding including grant number, donations of experimental material (drugs) and technical supports (language help, writing assistance or proof reading the article, analysis, etc) should be acknowledged. The disclosure of any conflict of interest or none is equally important when reporting. A *Conflict of Interest* is defined as a set of conditions in which professional judgment concerning a primary interest, such as the validity of research, may be influenced by a secondary interest, such as financial gain. A declaration of Conflict of Interest is a notification from the author that there is no financial, personal interest or belief that could affect the objective of the study. In case of the existence of conflict of interest, author should state the source and nature of that conflict.

## Referencing: Citation and List

Every academic writing requires appropriate reference *i.e.* the sources that were consulted in the preparation of report. The citation which includes quoting from other writers should be accompanied by respective names of the writer, titles and year of the publications. These information could facilitate tracking of the work. The citations provides credibility, add value and confer authority to reports. Citations show how up-to-date a report is. Hence, it is advisable to keep in touch with the most recent developments. Citations are a means of valuing intellectual capacity of authors of previous work and give room for prospective readers to crosscheck current report with the previous ones. Idea can be cited by changing the word order, summarize lengthy material or repeat the authors style of expression along with references. Author should be consistent in their style of citations within text and on the reference lists. Citation within text often include the name of author and year of publication. Informations including name of the author(s), year of publication, title of the publication, place of publication, publisher, volume, issue number, and page numbers should be provided in the reference lists to facilitate retrieval of report and enables reader to find the original source. For example, *Fajemiroye, 2000* within the text gives an idea that the work being cited was realized by Fajemiroye alone in the year 2000. This information could help reader to appreciate the depth of literature review and how current the author of the material is. In the case of *Fajemiroye et al., 2000*, there are indication that Fajemiroye reported a different work with his collaborators in the same year. However, ifit were to be *Fajemiroye* alone in the same year, author will have to introduce numbers, symbol or alphabet after the year to indicate different work of the same author in the same year (*Fajemiroye, 2000a; Fajemiroye, 2000b*). Citation of several authors with similar view or result on the matter of interest should be reported in a parenthesis with semi colon or any other appropriate apostrophe to separate each authors (*e.g.* Fajemiroye *et al.,* 2000; James *et al.,* 2008). However, the citation within the text could not provide information on the full name of the author (s), title of the work, edition, volume, issue, the page numbers of the works, the size of the work (book, article or reports) being cited, the journal or medium of this publication among others, hence, the introduction of "References" or "Bibliography" lists that have all these information at the end of the report is very important. Author should ensure that information being provided in the references are correct as incorrect surnames, journal/book titles, publication year and pagination may prevent access. When copying references from other sources, authors should be careful as they may already contain errors which could be transferred to the new reference lists. Irrespective of who committed the first error, the last author of a wrongful citation will always be held responsible. In case of newly released report without volume, issue and pagination, the use of the object identifier - DOI is encouraged. A DOI which is considered as a permanent link (*e.g.* http://dx.doi.org/10.1/5120000CA.) can be used to cite and link to electronic articles where an article is in-press and full citation details are not yet known, but the article is available online. Citation of a reference as '*in press*' implies that the item has been accepted for publication. However, such citations should be consistent or in the same style as all other references in the paper. The list of references is arranged alphabetically by surname of the author in order to enables the reader locate details of all the sources cited easily. Following the surname of author, the initials of other names may be abbreviated. It is a convention that the scientific names of plants within the titles of citation be italicized or underlined. Any citation to an electronic source should conform to your chosen citation style. It is important to note that different style of citation and reference listing could be established for different report; sometimes, citation could be arranged in numerical order, title could be bold, sources are sometimes italicized; hence, the most important aspect of referencing is to understand the required style and be consistent. Author should ensure that every reference cited in the text is also present in the reference list (and vice versa).

## 10.3 Supplementary Material

Sometimes, the author of a report could be limited by the number of words, figures, tables, schemes or references that are allowed in the main report, hence, supplementary material may remain the only option to provide additional informations to enhance scientific research. Supplementary files could provide space to publish supporting materialss, high-resolution images, background datasets, sound clips, etc. The material should be provided with a concise and descriptive caption. Video material and animation sequences should be provided with appropriate links.

In summary, the entire report should consistently feature correct use of scientific names (genus and species) in italicized or underlined form, appropriate units for parameters under measurement, correct language (simple and grammatically correct sentence) and technical terms, good and purposeful paragraphs, impersonal sentence (instead of saying, "We found that..." write, "it was found...") devoid of informal words, slang and excessive word contractions. Finally, author should read the report again and again to check all kinds of errors or ambiguity before seeking the help of professional reviewers.

# References

Abdillahi H S, Van Staden J, 2013. Application of medicinal plants in maternal healthcare and infertility: A South African Perspective. Planta Medica 79, 591–599.

Abo KA, Fred-Jaiyesimi AA, Jaiyesimi AEA, 2008. Ethnobotanical studies of medicinal plants used in the management of diabetes mellitus in South Western Nigeria. Journal of Ethnopharmacology 115(1), 67–71.

Aboaba O, Smith, SY, Olude F, 2006. Antibacterial effect of Edible Plant Extract on E coli 0157: h7. Pakistan Journal of Nutrition 5(4), 325–327.

Abondo A, Mbenkum F, Thomas D, Ethnobotany and the medicinal plants of the Korup rainforest project area, Cameroon. Proceedings of the International Conference on Medicinal Plants organised by Ministry of Health, Tanzania. 1991, pp. 391.

Abubakar MS, Musa AM, Ahmed A, Hussaini IM, 2007. The perception and practice of traditional medicine in the treatment of cancers and inflammations by the Hausa and Fulani tribes of Northern Nigeria. Journal of Ethnopharmacology 111(3), 625–629.

Abubakar MS, Sule MI, Pateh UU, Abdurahman EM, Haruna AK, Jahun BM, 2000. *In vitro* snake venom detoxifying action of the leaf extract of *Guiera senegalensis*. Journal of Ethnopharmacology 69(3), 253–257.

Abubakar S, Usman AB, Ismaila IZ, Aruwa G, Azizat SG, Ogbadu GH, Onyenekwe PC, 2014. Nutritional and Pharmacological Potentials of *Leptadenia Hastata* (Pers.) Decne. Ethanolic Leaves Extract. Journal of Food and Nutrition Research 2(1), 51–55.

Acharya E, Pokhrel B, 2007. Ethno-Medicinal Plants Used by Bantar of Bhaudaha, Morang, Nepal. Our Nature 4(1), 96–103.

Adam MS, and Allan VK, 2014. Anxiolytic drug discovery: what are the novel approaches and how can we improve them? Expert Opinion on Drug Discovery 9, 15-26.

Adamu HM, Abayeh OJ, Agho MO, Abdullahi AL, Uba A, Dukku HU, Wufem BM, 2005. An ethnobotanical survey of Bauchi State herbal plants and their antimicrobial activity. Journal of Ethnopharmacology 99(1), 1–4.

Adebayo AH, John-africa LB, Agbafor AG, Omotosho OE, Mosaku TO, 2014.

Aguilar ML, 2003. Etnomedicina en Mesoamérica, Arqueología Mexicana, Vol. X, No. 59, enero-febrero, INAH, México, pp. 26-31.

Álvarez Asomoza C, 2003. Los Hongos Sagrados de Teotenango, Estado de México, Arqueología Mexicana, Vol. X, No. 59, enero-febrero, INAH, México, pp. 38-41.

Antinociceptive and anti-inflammatory activities of extract of *Anchomanes difformis* in rats. Pakistan Journal of Pharmaceutical Sciences 27(2), 265–270.

Adebisi IM, Alebiosu OC, 2014. A Survey of Herbal Abortificients and Contraceptives in Sokoto, North-West Nigeria. International Journal of Current Research in Chemistry and Pharmaceutical Sciences 1(7), 81–87.

Adedapo AA, Abatan MO, Olorunsogo OO, 2004. Toxic effects of some plants in the genus Euphorbia on haematological and biochemical parameters of rats. Veterinarski Archive 74(1), 53–62.

Adelanwa EB, Tijjani AA, 2013. An ethno-medical survey of the flora of Kumbotso local government area of Kano State. Nigerian Journal of Pharmaceutical Science 12(1), 1–9.

Aderibigbe AO, Emudianughe TS, Lawal BAS, 2001. Evaluation of the antidiabetic action of *Mangifera indica* in mice. Phytotherapy Research 15(5), 456–458.

Adnan M, Jan S, Mussarat S, Tariq A, Begum S, Afroz A, Shinwari ZK, 2014. A review on ethnobotany, phytochemistry and pharmacology of plant genus *Caralluma* R.Br. The Journal of Pharmacy and Pharmacology 66, 1–18.

Adoum OA, Nenge HP, Chedi B, 2012. The Steroidal Component and Hypoglycaemic Effect of Stem Bark Extract of *Mitragyana inermis* (Willd) O. Kuntze (Rubiaceae) in Alloxan induced Diabetic Wistar rats. International Journal of Applied Biology and Pharmaceutical Technology 3(2), 169–174.

Adu F, Gbedema SY, Akanwariwiak WG, Annan K, Boamah VE, 2011. The Effects of *Acanthospermum hispidum* extract on the antibacterial activity of amoxicillin and ciprofloxacin. Hygeia Journal of Drugs and Medicine 3(1), 58–63.

Agaie BM, Salisu A, Ebbo AA, 2007. A survey of common toxic plants of livestock in Sokoto State, Nigeria. Scientific Research and Essay 2, 40–42.

Agarwal N, Dey CD, 1977. Behavioural and lethal effects of alcoholic extracts of *Evolvulus alsinoides* in albino mice. Journal of Physiology and Allied Sciences 31(2), 8.

Agbogidi OM, Akparobi SO, Eruotor PG, 2013. Health and environmental benefits of *Jatropha curcas* Linn. Applied Science Reports 1(2), 36–39.

Agunu A, Yusuf S, Andrew GO, Zezi AU, Abdurahman EM, 2005. Evaluation of five medicinal plants used in diarrhoea treatment in Nigeria. Journal of Ethnopharmacology 101(1-3), 27–30.

Ahmed AS, Mcgaw LJ, Moodley N, Naidoo V, Eloff JN, 2014. Phenolic composition of *Ozoroa* and *Searsia* species (Anacardiaceae) used in South African traditional medicine for treating diarrhoea. South African Journal of Botany 95, 9–18.

Ahmed MK, Mabrouk MA, Anuka JA, Attahir A, Tanko Y, Wawat AU, Yusuf MS, 2010. Studies of the Effect of Methanolic Stem Bark Extract of *Lannea acida* on Fertility and Testosterone in Male Wistar Rats. Asian Journal of Medical Sciences 2(6), 253–258.

Ahuja SS, 2007. Assuring quality of drugs by monitoring impurities. Adv Drug Deliv Rev 59, 3–11.

Ajaiyeoba EO, Oladepo O, Fawole OI, Bolaji OM, Akinboye DO, Ogundahunsi OAT, Oduola AMJ, 2003. Cultural categorization of febrile illnesses in correlation with herbal remedies used for treatment in Southwestern Nigeria. Journal of Ethnopharmacology 85(2-3), 179–185.

Akah PA, Njike HA, 1990. Some pharmacological effects of rhizome aqueous extract of Anchomanes difformis. Fitoterapia 61(4), 368–370.

Akah PA, Nwambie AI, 1994. Evaluation of Nigerian traditional medicines: 1. Plants used for rheumatic (inflammatory) disorders. Journal of Ethnopharmacology 42(3), 179–182.

Akhtar SS, 2013. Evaluation of Cardiovascular Effects of *Citrus aurantifolia* (Linn.) Fruit. Social Science Research Network SSRN http://papers.ssrn.com/sol3/papers. cfm?abstract_id=2279447.

Akhtar MS, Iqbal J, 1991. Journal of Ethnopharmacology 31(1), 49-57.

Alagesaboopathi C, 2009. Ethnomedicinal plants and their utilization by villagers in Kumaragiri Hills of Salem district of Tamilnadu, India. African Journal of Traditional, Complementary and Alternative Medicines 6(3), 222–227.

Aleixo R, Azevedo B De, Biserra Y, Carneiro E, Neto DF, Fatima M De, Coelho B, 2014. Uses of medicinal plants in Rio Grande Do Norte, Journal of Global Bioscience 3(4), 749–762.

Al-Hashem FH, 2010. Gastroprotective effects of aqueous extract of Chamomilla recutita against ethanol-induced gastric ulcers. Saudi Med. J. 31, 1211–1216.

Aliero AA, Wara SH, 2009. Validating the medicinal potential of *Leptadenia hastata*. African Journal of Pharmacognosy and Pharmacology 3(1), 335–338.

Ali-Shtayeh MS, Jamous RM, Jamous RM, 2015. Plants used during pregnancy, childbirth, postpartum and infant healthcare in Palestine. Complementary Therapies in Clinical Practice 21(2), 84–93.

Aliyu AB, Ibrahim H, Musa AM, Ibrahim MA, Oyewale AO, Amupitan JO, 2010. *In vitro* evaluation of antioxidant activity of *Anisopus mannii* N.E.Br. Journal of Biotechnology 9(16), 2437–2441.

Aliyu AB, Ibrahim MA, Musa AM, Musa AO, Kiplimo JJ, Oyewale AO, 2013. Free radical scavenging and total antioxidant capacity of root extracts of *Anchomanes difformis* Engl. (Araceae). Acta Poloniae Pharmaceutica - Drug Research 70(1), 115–121.

Aliyu AB, Musa AM, Abdullahi MS, Ibrahim MA, Tijjani MB, Aliyu MS, Oyewale AO, 2011. Activity of saponin fraction of *Anisopus mannii* against some pathogenic microorganisms. Journal of Medicinal Plants Research 5(31), 6709–6713.

Aliyu AB, Musa AM, Ibrahim H, Oyewale AO, 2009. Preliminary phytochemical screening and antioxidant activity of leave extract of *Albizia chevalieri* Harms. Bayero Journal of Pure and Applied Sciences 2(1), 149–153.

Alpsoy L, Sahin H, Karaman S, 2011. Anti-oxidative and anti-genotoxic effects of methanolic extract of Mentha pulegium on human lymphocyte culture. Toxicol. Ind. Health 27, 647–654.

Amos S, Kolawole E, Akah P, Wambebe C, Gamaniel K, 2001. Behavioral effects of the aqueous extract of *Guiera senegalensis* in mice and rats. International Journal of Phytotherapy and Phytopharmacology 8(5), 356–361.

Amri E, Kisangau DP, 2012. Ethnomedicinal study of plants used in villages around Kimboza forest reserve in Morogoro, Tanzania. Journal of Ethnobiology and Ethnomedicine 8(1), 1.

Amrita B, Kumar PU, Suchi D, Arvind J, Bhusan SH, 2012. Hair Growth Potential of *Evolvulus alsinoides* Linn. Plant Extract in Albino Rats. International Research Journal of Pharmacy 3(5), 314–319.

Ancolio C, Azas N, Mahiou V, Ollivier E, Di Giorgio C, Keita A, Balansard G, 2002. Antimalarial activity of extracts and alkaloids isolated from six plants used in traditional medicine in Mali and Sao Tome. Phytotherapy Research 16(7), 646–649. http://doi.org/10.1002/ptr.1025

Andrade DM, Reis CDF, Castro PFDS, Borges LL, Amaral NO, Torres IMS, Lavorenti Rocha M, 2015. Vasorelaxant and hypotensive effects of jaboticaba fruit (Myrciaria cauliflora) extract in rats. *Evidence-Based Complementary and Alternative Medicine 2015*, 1-8.

Anwar F, Latif S, Ashraf M, Gilani AH, 2007. *Moringa oleifera*: a food plant with multiple medicinal uses. Phytotherapy Research 21(1), 17–25.

Anywar G, Oryem-origa H, Kamatenesi-mugisha M, 2014. Antibacterial and Antifungal Properties of Some Wild Nutraceutical Plant Species from Nebbi District, Uganda. British Journal of Pharmaceutical Research 4(14), 1753–1761.

Anzenbacher P and Anzenbacherova,. 2001. Cytochromes P450 and metabolism of xenobiotics. Cell.Mol. Life Sci. 58, 737–747.

Aquino R, Peluso G, De Tommasi N, De Simone F, Pizza C, 1996. New polyoxypregnane ester derivatives from *Leptadenia hastata*. Journal of Natural Products 59(6), 555–564.

Araújo E de L, Randau KP, Sena-Filho JG, Pimentel RMM, Xavier HS, 2008. Acanthospermum hispidum DC (Asteraceae): perspectives for a phytotherapeutic product. *Revista Brasileira de Farmacognosia 18*, 777–784.

Asase A, Kokubun T, Grayer RJ, Kite G, Simmonds MS, Alfre AO, Ogunayo GT, 2008. Chemical constituents and antimicrobial activity of medicinal plants from Ghana: *Cassia sieberiana, Haematostaphis barteri, Mitragyna inermis* and *Pseudocedrela kotschyi*. Phytotherapy Research 22, 1013–1016.

Asase A, Oteng-Yeboah AA, Odamtten GT, Simmonds MSJ, 2005. Ethnobotanical study of some Ghanaian anti-malarial plants. Journal of Ethnopharmacology 99(2), 273–9. http://doi.org/10.1016/j.jep.2005.02.020

Asgary S, Sahebkar A, Afshani MR, Keshvari M, Haghjooyjavanmard S, Rafieian Kopaei M, 2014. Clinical evaluation of blood pressure lowering, endothelial function improving, hypolipidemic and anti-inflammatory effects of pomegranate juice in hypertensive subjects. *Phytotherapy Research/: PTR 28*(2), 193–9.

Asha V, Jeeva S, Paulraj K, 2014. Phytochemical and FT-IR spectral analysis of Caralluma geniculata Grev. et Myur. an endemic medicinal plant. Journal of Chemical and Pharmaceutical Research 6(7), 2083–2088.

Ataman EJ, Idu M, 2015. Renal effects of *Anchomanes difformis* crude extract in wistar rats. Avicenna Journal of Phytomedicine 5(1), 17–25.

Atawodi SE, Olowoniyi OD, Obari MA, Ogaba I, 2014. Ethnomedical Survey of Adavi and Ajaokuta Local Government Areas of Ebiraland, Kogi. Annual Research and Review in Biology 4(24), 4344–4360.

Atawodi SEO, Onaolapo GS, 2010. Comparative *in vitro* antioxidant potential of different parts of *Ipomoea asarifolia*, Roemer and Schultes, *Guiera senegalensis*, J.F. Gmel and *Anisopus mannii* N.E. Brown. Brazilian Journal of Pharmaceutical Sciences 46(2), 245–250.

Atif A, 2012. *Acacia nilotica*: A plant of multipurpose medicinal uses. Journal of Medicinal Plants Research 6(9), 1492–1496.

Atkinson AJ, *Principles of Clinical Pharmacology*. San Diego, CA: Academic Press; Elsevier. 2012.

Attah AF, O'Brien M, Koehbach J, Sonibare MA, Moody JO, Smith TJ, Gruber CW, 2012. Uterine contractility of plants used to facilitate childbirth in Nigerian ethnomedicine. Journal of Ethnopharmacology 143(1), 377–82.

Austarheim I, Nergard CS, Sanogo R, Diallo D, Paulsen BS, 2012. Inulin-rich fractions from *Vernonia kotschyana* roots have anti-ulcer activity. Journal of Ethnopharmacology 144(1), 82–85.

Austin DF, 2008. *Evolvulus alsinoides* (Convolvulaceae): An American herb in the Old World. Journal of Ethnopharmacology 117(2), 185–198.

Awobajo FO, Olatunji-Bello II, Ogbewey LI, 2013. Reproductive impact of aqueous leaf extract of *Mangifera indica* Mango on some reproductive functions in female Sprague-Dawley rats. Biology and Medicine 5(1), 58–64.

Awodele O, Oreagba IA, Odoma S, da Silva JAT, Osunkalu VO, 2012. Toxicological evaluation of the aqueous leaf extract of *Moringa oleifera* Lam. (Moringaceae). Journal of Ethnopharmacology 139(2), 330–336.

Ayyanar M, Ignacimuthu S, 2009. Herbal medicines for wound healing among tribal people in Southern India: Ethnobotanical and scientific evidences. International Journal of Applied Research in Natural Products 2(3), 29–42.

Bagrov AY, Roukoyatkina NI, Pinaev AG, Dmitrieva RI, Fedorova OV, 1995. Effects of two endogenous Na+,K(+)-ATPase inhibitors, marinobufagenin and ouabain, on isolated rat aorta. *European Journal of Pharmacology 274*(1-3), 151–8.

Bah S, Diallo D, Dembélé S, Paulsen BS, 2006. Ethnopharmacological survey of plants used for the treatment of schistosomiasis in Niono District, Mali. Journal of Ethnopharmacology 105(3), 387–399.

Baharvand-Ahmadi B, Bahmani M, Zargaran A, 2016. A brief report of Rhazes manuscripts in the field of cardiology and cardiovascular diseases. *International Journal of Cardiology 207*(1), 190–191.

Bailey DG, Bend JR, Arnold JMO, Tran LT, Spence JD, 1996. Erythromycin-felodipine interaction: magnitude, mechanism, and comparison with grapefruit juice. Clin Pharmacol Ther. 60:25–33.

Bailey DG, Arnold JMO, Bend JR, Tran LT, Spence JD, 1995. Grapefruit juice-felodipine interaction: reproducibility and characterization with the extended release drug formulation. Br J Clin Pharmacol. 40:135–140.

Bailey DG, Arnold JMO, Munoz C, Spence JD, 1993. Grapefruit juice-felodipine interaction: mechanism, predictability and effect of naringin. Clin Pharmacol Ther. 53:637–642.

Bailey DG, Munoz C, Arnold JMO, Strong HA, Spence JD, 1992. Grapefruit juice and naringin interaction with nitrendipine. Clin Pharmacol Ther. 51:156.

Bailey DG, Spence JD, Munoz C, Arnold JMO, 1991. Interaction of citrus juices with felodipine and nifedipine. Lancet 337:268–269.

Bakhita A, Adam SEI, 1978. Toxicity of *Acanthospermus hispidum* to mice. Journal of Complementary Pathology *88*, 443–448.

Bako Y, Ibrahim M, Mohammad S, Zubairu M, Bulus T, 2014. Toxicity studies of aqueous, methanolic and hexane leaf extracts of *Guiera senegalensis* in rats. International Journal of Scientific and Engineering Research 5(10), 1338–1347.

Balaji R, Suba V, Rekha N, Deecaraman M, 2009. Hepatoprotective activity of methanolic fraction of *Jatropha curcas* on aflatoxin b1 induced hepatic carcinoma. International Journal of Physical Sciences 2, 287–296.

Balangcod TD, Balangcod AKD, 2011. Ethnomedical knowledge of plants and healthcare practices among the Kalanguya tribe in Tinoc, Ifugao, Luzon, Philippines. Indian Journal of Traditional Knowledge 10(2), 227–238.

Balde MA, Traore MS, Diane S, Diallo MST, Tounkara TM, Camara A, Balde AM, 2015. Ethnobotanical survey of medicinal plants traditionally used in Low and Middle - Guinea for the treatment of skin diseases. Journal of Plant Sciences 3, 32–39.

Bapna S, Ramaiya M, Chowdhary A, 2014. Brine shrimp toxicity and invitro antimalarial activity of *Citrus aurantifolia* (Christm.) Swingle against *Plasmodium falciparum*. Journal of Pharmacy and Biological Science 9(5), 24–27.

Baradaran A, Nasri H, Rafieian-Kopaei M, 2014. Oxidative stress and hypertension: Possibility of hypertension therapy with antioxidants. *Journal of Research in Medical Sciences/: The Official Journal of Isfahan University of Medical Sciences* 19(4), 358-67.

Barbosa LC, Dias de Morais M, de Paula CA, da Silva Ferreira MC, Jordao AA, Andrade e Silva ML, *et al.*, 2012. Mikania glomerata Sprengel (Asteraceae) influences the mutagenicity induced by doxorubicin without altering liver lipid peroxidation or antioxidant levels. J. Toxicol. Environ. Health Part A 75, 1102–1109.

Baxter K, and Stockley IH, Stockley's Drug Interactions: A Source Book of Interactions,their Mechanisms, Clinical Importance, and Management. London; Chicago: Pharmaceutical Press. 2008.

Bayala B, Rubio-Pellicer MT, Zongo M, Malpaux B, Sawadogo L, 2011. Activité anti-androgénique de *Leptadenia hastata* (Pers.) Decne: Effet compétitif des extraits aqueux de la plante et du propionate de testostérone sur des rats impubères castrés. Biotechnology, Agronomy and Society and Environment 15(2), 223–229.

Bayala B, Telefo PB, Bassole IHN, Tamboura HH, Belemtouri RG, Sawadogi L, Dacheux JL, 2011. Anti-sprmatogenic Activity of *Leptadenia hastata* (Pers.) Decne Leaf Stems Aqueous Extracts in Male Wistar Rats. Journal of Pharmacology and Toxicology 6(4), 391–399.

Bayala B, Téléfo PB, Savadogo A, Sawadogo L, Malpaux B, 2012. Combined effects of testosterone propionate and *Leptadenia hastata* Pers. (Decne) aqueous extracts on immature castrated male rats. Journal of Medicinal Plants Research 6(15), 2925–2931.

Begum SN, Ravikumar K, Ved DK, 2014. 'Asoka' – an important medicinal plant, its market scenario and conservation measures in India. Curr. Sci. 107, 26–28.

Belem B, Nacoulma B, Gbangou R, Kambou S, Hansen HH, Gausset Q, Boussim IJ, 2007. Use of non wood forest products by local people bordering the" Parc National Kaboré Tambi", Burkina Faso. Journal for Transdisciplinary Environmental Studies 6(1), 18.

Bello A, Aliero AA, Saidu Y, Muhammad S, 2011a. Hypoglycaemic and hypolipidaemic effects of *Leptadenia hastata* (Pers.) Decne in Alloxan induced diabetic rat. Nigerian Journal of Basic and Applied Science 19(2), 187–192.

Bello A, Aliero AA, Saidu Y, Muhammad S, 2011b. Phytochemical screening, polyphenolic content and alpha-glucosidase inhibitory potential of *Leptadenia hastata* (Pers.) Decne. Nigerian Journal of Basic and Applied Science 19(2), 181–186.

Benítez F, Los Indios de México. Los hongos alucinantes, Ed. Era, México. 1964.

Ben-nasr H, Ali M, Abderrahim B, Salama M, Ksouda K, Zeghal K, 2013. Potential Phytotherapy use of *Artemisia* Plants: Insight for anti-hypertension. Journal of Applied Pharmaceutical Science 3(05), 120–125.

Berka K, Hendrychova T, Anzenbacher P, and Otyepka M, 2011. Membrane position of ibuprofen agrees with suggested access path entrance to cytochrome P450 2C9 activesite. J. Phys. Chem. A 115, 11248–11255.doi:10.1021/jp204488j

Betti JL, Rostand S, Yemefa M, Tarla FN, 2011. Contribution to the knowledge of non wood forest products of the far north region of Cameroon/: Medicinal plants sold in the Kousséri market. Journal of Ecology and the Natural Environment 3, 241–254.

Bhandary MJ, Chandrasekhar KR, Kaveriappa KM, 1995. Medical ethnobotany of the Siddis of Uttara Kannada District, Karnataka, India. J. Ethnopharmacol. 47, 149–158. 19.

Bhatt V, Karakoti R, Bhatt B, Singh AK, Sharma DK, Vidyapeeth D, 2015. Formulation and evaluation of antiemetic nasal. World Journal of Pharmacy and Pharmaceutical Sciences 4(01), 1381–1392.

Bhattacharjee R, Sil PC, 2006. The protein fraction of Phyllanthus niruri plays a protective role against acetaminophen induced hepatic disorder via its antioxidant properties. Phytother. Res. 20, 595–601

Bian K, Toda N, 1988. Effects of sophoramine, an alkaloid from Sophora alopecuroides on isolated dog blood vessels. *Journal of Ethnopharmacology* 24(2-3), 167–78.

Biljana BP, 2012. Historical review of medicinal plants' usage. Pharmacogn Rev. 6, 1-5.

Biswas TK and Debnath P, 1972. Asoka (Saraca indica Linn) – a cultural and scientific evaluation. Indian J. Hist. Sci. 7, 99–114.

Boareto AC, Muller JC, Bufalo AC, Botelho GGK., Araujo, SL De, Ann M, Dalsenter PR, 2008. Toxicity of artemisinin [*Artemisia annua* L.] in two different periods of pregnancy in Wistar rats. Reproductive Toxicology 25, 239–246.

Borokini TI, Ighere DA, Clement M, Ajiboye TO, Alowonle AA, 2013. Journal of Medicinal Plants Studies Ethnobiological survey of traditional medicine practice for Women' s health in Oyo State. Journal of Medicinal Plants Studies 1, 17–29.

Borokini TI, Ighere DA, Clement M, Ajiboye TO, Alowonle AA, 2013. Ethnobiological survey of traditional medicine practice for Women's health in Oyo State. Journal of Medicinal Plants 1(5).

Both FL *et al.*, 2005. Psychopharmacological profile of the alkaloid psychollatine as a 5HT2A/C serotonin modulator. J Nat Prod 68, 374–80.

Bourin M and Hascoët M, 2003. The mouse light/dark box test. European Journal of Pharmacology 463, 55–65.

Bouskela E, Cyrino FZ, Marcelon G, 1994a. Possible mechanisms for the inhibitory effect of Ruscus extract on increased microvascular permeability induced by histamine in hamster cheek pouch. *Journal of Cardiovascular Pharmacology* 24(2), 281–5.

Bouskela E, Cyrino FZ, Marcelon G, 1994b. Possible mechanisms for the venular constriction elicited by Ruscus extract on hamster cheek pouch. *Journal of Cardiovascular Pharmacology* 24(1), 165–70.

Braquet P, Hosford D, 1991. Ethnopharmacology and the development of natural PAF antagonists as therapeutic agents. *Journal of Ethnopharmacology* 32(1-3),

Breithaupt-Grögler K, Ling M, Boudoulas H, Belz GG, 1997. Protective effect of chronic garlic intake on elastic properties of aorta in the elderly. *Circulation* 96(8), 2649–55.

Bridson D, and Forman L, The Herbarium Handbook. Royal Botanic Gardens, Kew. 3rd Edn (Edi.), 1999, pp: 4.

British Columbia Ministry of Forests. Techniques and procedures for collecting, preserving, processing, and storing botanical specimens. Res. Br., B.C. Min. For., Victoria, B.C. Work. Pap. 18/1996.

Budzinski JW, Foster BC, Vandenhoek S, Arnason JT, 2000. An *in vitro* evaluation of human cytochrome P450 3A4 inhibition by selected commercial herbal extracts and tinctures. Phytomedicine 7, 273–282.

Buening MK, Change RL, Huang MT, Fortner JG, Wood AW, Conney AH. 1981. Activation and inhibition of benzo(a)pyrene and aflatoxin B1 metabolism in human liver microsomes by naturally occurring flavonoids.Cancer Res. 41:67–72.

Bueno NR, Castilho RO, Costa RB, Pott A, Pott VJ, Scheidt GN, Batista MS, 2005. Medicinal plants used by the Kaiowá and Guarani indigenous populations in the Caarapó Reserve, Mato Grosso do Sul, Brazil. Acta Botânica Brasileira 19 (1), 39-44.

Burkill HM, 1995. *The Useful Plants of West Tropical Africa. Families J-L*, (Vol. 3.). Kew: Royal Botanical Gardens.

Burkill HM, 1985. The useful plants of West Tropical Africa. Royal Botanic Gardens 1(5), 33.

Burque RK, Francesconi LP, Victorino AT, Mascarenhas MÁ, Ceresér KM, 2015. Determinação de compostos fenólicos e avaliação da atividade antioxidante de *Lafoensia pacari* (LYTHRACEAE). Revista Eletrônica de Farmácia 12(1), 1-10.

Bussmann RW, 2006. Ethnobotany of the Samburu of Mt. Nyiru, South Turkana, Kenya. Journal of Ethnobiology and Ethnomedicine 10, 1–10.

Butterweck V *et al.*, 2002. In vitro receptor screening of pure constituents of St. John's wort reveals novel interactions with a number of GPCRs. Psychopharmacology (Berl) 162, 193 – 202.

Bye R, and Linares E, 1999. Plantas Medicinales Prehispánicas, Arqueología Mexicana, Vol. VII, No. 39, septiembre-octubre, INAH, México, pp. 4-6.

Cabral PRF, Pasa, MC, 2009. Mangava-brava: *Lafoensia pacari* A. St. - Hil. (Lythraceae) e a etnobotânica em Cuiabá, MT. Revista Biodiversidade 8, 1.

Cáceres, A., Freire, V., Girón, L. M., Avilés, O., Pacheco, G. 1991. *Moringa oleifera* (Moringaceae): ethnobotanical studies in Guatemala. Economic Botany, 45(4), 522–523.

Cafferata, L.F.R., Gatti, W.O., Mijailosky, S., Inibiolp, I.B., Médicas, F.D.C. 2010. Secondary gaseous metabolites analyses of wild *Artemisia annua* L. Molecular Medicinal Chemistry, 21, 44–47.

Cárdenas R, El libro de la Mitología. Historias, Leyendas y Creencias Mágicas Obtenidas de la Tradición Oral, Atelí y Cío Ltda., Punta Arenas, Chile. 1997.

Carlini EA and Burgos V, 1979. Screening farmacológico de ansiolíticos: Metodologia laboratorial e comparação entre diazepam e clorobenzepam. Rev Assoc Med Bras 1, 25–31.

Carvalho de Lucena, K.F., Rodrigues, J.M.N., Campos, É.M., Dantas, A.F.M., Pfister, J.A., Cook, D., Riet-Correa, F. 2014. Poisoning by *Ipomoea asarifolia* in lambs by the ingestion of milk from ewes that ingest the plant. Toxicon, 92, 129–132.

Cauli O and Morelli M, 2005. Caffeine and the dopaminergic system. Behav Pharmacol 16, 63–77.

Caver, S., Maksimovi, M., Vidic, D., Paric, A. 2012. Chemical composition and antioxidant and antimicrobial activity of essential oil of *Artemisia annua* L. from Bosnia. Industrial Crops and Products, 37, 479–485.

Cechinel Filho V, Yunes RA, 1998. Estratégias para a obtenção de compostos farmacologicamente ativos a partir de plantas medicinais: conceitos sobre modificação estrutural para otimização da atividade. Química Nova 21(1), 99–105.

Chabi, S.K., Sina, H., Adoukonou-Sagbadja, H., Ahoton, L.E., Roko, G.O., Saidou, A., Baba-Moussa, L. (2014). Antimicrobial activity of *Anacardium occidentale* L. leaves and barks extracts on pathogenic bacteria. African Journal of Microbiology Research, 8(25), 2458–2467.

Chakraborty, A.K., Gaikwad, A.V, Singh, K.B. 2012. Phytopharmacological review on *Acanthospermum hispidum*. Journal of Applied Pharmaceutical Science, 2(1), 144–148.

Chakraborty AK, Gaikwad AV, Singh KB, 2012. Phytopharmacological review on Acanthospermum Hispidum. *Journal of Applied Pharmaceutical Science* 2(1), 144–148.

Chan K, Lo AC, Yeung JH, Woo KS, 1995. The effects of Danshen (Salvia miltiorrhiza) on warfarin pharmacodynamics and pharmacokinetics of warfarin enantiomers in rats. *The Journal of Pharmacy and Pharmacology* 47(5), 402–6.

Chavkin C *et al.*, 2004. Salvinorin A, an active component of the hallucinogenic sage Salvia divinorum, is a highly efficacious kappa opioid receptor agonist: structural and functional considerations. J Pharmacol Exp Ther 308, 1197-1203.

Chen CX, Jin RM, Li YK, Zhong J, Yue L, Chen SC, Zhou JY, 1992. Inhibitory effect of rhynchophylline on platelet aggregation and thrombosis. *Zhongguo Yao Li Xue Bao = Acta Pharmacologica Sinica* 13(2), 126–30.

Chen XJ, Fan HY, Yao YF, Zhang JX, Gu WX, Liu XM, Liu JF, 1987. Relation of the antihypertensive effect and central alpha-adrenoceptor of rhomotoxin. *Zhongguo Yao Li Xue Bao = Acta Pharmacologica Sinica* 8(3), 247–50.

Chen Z, Sun J, Chen H, *et al.*, 2010. Comparative pharmacokinetics and bioavailability studies of quercetin, kaempferol and isorhamnetin after oral administration of Ginkgo biloba extracts, Ginkgo biloba extract phospholipid complexes and Ginkgo biloba extract solid dispersions in rats. Fitoterapia 81: 1045 e 1052.

Cheung K, Hinds JA, Duffy P, 1989. Detection of poisoning by plant-origin cardiac glycoside with the Abbott TDx analyzer. *Clinical Chemistry* 35(2), 295–7.

Chitemerere, T.A, Mukanganyama, S. 2011. *In vitro* antibacterial activity of selected medicinal plants from Zimbabwe. The African Journal of Plant Science and Biotechnology, 5(1), 1–7.

Chiou WF, Shum AY, Liao JF, Chen CF, 1997. Studies of the cellular mechanisms underlying the vasorelaxant effects of rutaecarpine, a bioactive component extracted from an herbal drug. Journal of Cardiovascular Pharmacology 29(4), 490–8.

Chunlaratthanaphorn, S., Lertprasertsuke, N., Srisawat, U., Thuppia, A., Ngamjariyawat, A., Suwanlikhid, N., Technol, S.J.S. 2007. Acute and subchronic toxicity study of the water extract from root of *Citrus aurantifolia* (Christ. et Panz.) Swingle in rats. Songklsnsksrin Journal of Science and Technology, 29(Suppl. 1), 125–139.

Cinelli *et al.*, 2007. Comparative analysis and physiological impact of different tissue biopsy methodologies used for the genotyping of laboratory mice. Lab Anim 41, 174-184.

Citarella L, Medicinas y Culturas en la Araucanía, Trafkin, Cooperación Italiana en Salud, Sudamericana, Santiago, Chile. 1995.

Connor JD *et al.*, 2002. Comparison of effects of khat extract and amphetamine on motor behaviors in mice. J Ethnopharmacol 81, 65–71.

Cotton CM, Ethnobotany. Principles and applications. John Wiley and Sons Ltd., Chichester. 1996.

Cox PA, and Balick MJ, 1994. The Ethnobotanical Approach to Drug Discovery, Scientific American, Vol. 270, No. 6, pp. 82-87.

Crawley JN and Goodwin FK, 1980. Preliminary report of a simple animal behavior for the anxiolytic effects of benzodiazepines. Pharmacology Biochemistry and Behavior 13, 167–170.

Cryan JF and Lucki I, 2000. Antidepressant-like behavioral effects mediated by 5-hydroxytryptamine(2C) receptors. J. Pharmacol. Exp. Ther., 295, 1120–112

Cunha LC *et al.*, 2009. Avaliação da toxicidade aguda e subagu-da, em ratos, do extrato etanólico das folhas e do látex de Synadenium umbellatum Pax. Rev. Bras. Farmacognosia 19, 403-411.

D'Aquila PS *et al.*, 1997. Antianhedonic actions of the novel serotonergic agent flibanserin, a potential rapidly-acting antidepressant. Eur J Pharmacol 340: 121–132.

Dada EO, Ekundayo FO, Makanjuola OO, 2014. Antibacterial activities of *Jatropha curcas* (Linn) on coliforms isolated from surface waters in Akure, Nigeria. International Journal of Biomedical Science 10(1), 25–30.

Daisy P and Rajathi M, 2009. Hypoglycemic Effects of Clitoria ternatea Linn. (Fabaceae) in Alloxan-induced Diabetes in Rats. Tropical Journal of Pharmaceutical Research 8 (5): 393-398.

Damasceno DC, Volpato GT, Calderon Ide M, Aguilar R, Rudge MV, 2004. Effect of Bauhinia forficata extract in diabetic pregnant rats: maternal repercussions. Phytomedicine 11, 196–201.

Dambatta SH, 2011. A survey of major ethno medicinal plants of Kano North, Nigeria, their knowledge and uses by traditional healers. Bayero Journal of Pure and Applied Sciences 4(2), 28–34.

Dassprakash MV, Arun R, Abraham SK, Premkumar K, 2012. *In vitro* and *in vivo* evaluation of antioxidant and antigenotoxic potential of *Punica granatum* leaf extract. Pharm. Biol. 50, 1523–1530.

De Leo M, De Tommasi N, Sanogo R, Autore G, Marzocco S, Pizza C, Braca A, 2005. New pregnane glycosides from *Caralluma dalzielii*. Steroids 70(9), 573–585.

Deeni YY, Hussain HSN, 1994. Screening of *Vernonia kotschyana* for antimicrobial activity and alkaloids. Pharmaceutical Biology 32(4), 388–395.

Deepa N, Rajendran NN, 2007. Anti-tumor Activity of *Acanthospermum hispidum* DC on Dalton Ascites Lymphoma in mice. Natural Product Sciences 13(3), 234–240.

Delaney TA, Donnelly AM, 1996. Garlic dermatitis. *The Australasian Journal of Dermatology* 37(2), 109–10.

Del Cerro S *et al.*, 1990. Inhibition of long-term potentiation by an antagonist of platelet-activating factor receptors. Behav Neural Biol. 54, 213- 217.

Deng C, *et al.*, 2007. Recent developments in sample preparation techniques for chromatography analysis of traditional Chinese medicines. J Chromatogr A. 1153, 90-6.

Deng X, Zheng S, Gao G, Fan G, Li F, 2008. Determination and pharmacokinetic study of indirubin in rat plasma by high-performance liquid chromatography. Phytomedicine 15:277e283.

Deree J *et al.*, 2008. "Insights into the regulation of TNF-alpha production in human mononuclear cells: the effects of non-specific phosphodiesterase inhibition". Clinics 63 321–8.

Dervendzi V, 1992. Contemporary treatment with medicinal plants. Skopje: Tabernakul 5–43.

Deslypere JP, Vermeulen A, 1991. Rhabdomyolysis and simvastatin. Ann Intern Med. 114:342.

Dhakar R, Pooniya B, Gupta M, Maurya S, Bairwa N, Sanwarmal, 2011. *Moringa/*: The herbal gold to combat malnutrition. Chronicles of Young Scientists 2(3), 119.

Diao X, Ma Z, Wang H, *et al.*, 2013. Simultaneous quantitation of 3-n-butylphthalide (NBP) and its four major metabolites in human plasma by LC-MS/MS using deuterated internal standards. J Pharm Biomed Anal. 78e79:19e26.

Díaz JL, 2003. Las Plantas Mágicas y la Conciencia Visionaria, Arqueología Mexicana, Vol. X, No. 59, enero-febrero, INAH, México, pp. 16-25.

Dickstein ES, Kunkel FW, 1980. Foxglove tea poisoning. The American Journal of Medicine 69(1), 167–9.

Dillehay T, Monte Verde. Un Asentamiento Humano del Plestoceno Tardío en el Sur de Chile, Ed. LOM, Universidad Austral de Chile, Santiago, Chile. 2004.

Dimo T, Rakotonirina SV, Tan PV, Azay J, Dongo E, Kamtchouing P, Cros G, 2007. Effect of *Sclerocarya birrea* (Anacardiaceae) stem bark methylene chloride/methanol extract on streptozotocin-diabetic rats. Journal of Ethnopharmacology 110(3), 434–438.

Diouf A, Cisse A, Gueye SS, Mendes V, Siby T, Diouf DR, Bassene E, 1999. Toxocological study of *Guiera senegalensis* Lam (Combretaceae). Dakar Medical 45(1), 89–94.

Diwani GE, Rafie SE, Hawash S, 2009. Antioxidant activity of extracts obtained from residues of nodes leaves stem and root of Egyptian *Jatropha curcas*. Journal of Pharmacy and Pharmacology 3(November), 521–530.

Doka IG, Yagi SM, 2009. Ethnobotanical Survey of Medicinal Plants in West Kordofan Western Sudan. Ethnobotanical Leaflets 13, 1409–1416.

Doly M, Droy-Lefaix MT, Braquet P, 1992. Oxidative stress in diabetic retina. *EXS* 62, 299–307.

Dresser, GK *et al.*, 2000. Pharmacokinetic-pharmacodynamic consequences and clinical relevance of cytochrome P450 3A4 inhibition. Clin. Pharmacokinet. 38, 41–57.

Dunham NW and Miya TS, 1957. A note on a simple apparatus for detecting neurological deficit in rats and mice. J Am Pharm Assoc, 46, 208–209.

Edgar B, Bailey DG, Bergstrand R, Johnsson G, Regardh CG. 1992. Acute effects of drinking grapefruit juice on the pharmacokinetics and pharmacodyanmics of felodipine–and its potential clinical relevance. Eur J Clin Pharmacol. 42:313–317.

El-Gied AAA, Abdelkareem AM, Hamedelniel EI, 2015. Investigation of cream and ointment on antimicrobial activity of *Mangifera indica* extract. Journal of Advanced Pharmaceutical Technology and Research 6(2), 53–57.

El-Hidayah TM, Abdulhadi NH, Badico EEM, Mohammed EY, 2011. Toxic potential of ethanolic extract of *Acacia nilotica* (Garad) in rats. Sudan Journal of Medical Science 6(1), 1–6.

Eliade M, El Chamanismo y las Técnicas Arcaicas del Éxtasis, Fondo de Cultura Económica, México. 2003.

EL-Kamali HH, 2009. Ethnopharmacology of medicinal plants used in North Kordofan (Western Sudan). Ethnobotanical Leaflets 2009(1), 24.

El-Kamali HH, El-Khalifa KF, 1999. Folk medicinal plants of riverside forests of the Southern Blue Nile district, Sudan. Fitoterapia 70(5), 493–497.

Ellena R, Quave CL, Pieroni A, 2012. Comparative medical ethnobotany of the senegalese community living in Turin (Northwestern Italy) and in Adeane (Southern Senegal). Evidence-Based Complementary and Alternative Medicine 2012(4).

Elrahman OF, Abuelgasim IA, Galal M, 2008. Toxicopathological effects of *Guiera senegalensis* extracts in wistar albino rats. Journal of Medicinal Plant Research 2(January), 1–4.

Emadi F, Sharifzadeh M, Yassa N, 2010. Investigation of Iranian *Artemisia annua* sedative effects in mice. Planta Medica 76(12), 55.

Eneojo AS, Egwari LO, Mosaku TO, 2011. *In vitro* Antimicrobial screening on *Anchomanes difformis* (Blume) Engl. leaves and rhizomes against selected pathogens of public health importance. Advances in Biological Research 5(4), 221–225.

Esimone CO, Nworu CS, Jackson CL, 2008. Cutaneous wound healing activity of a herbal ointment containing the leaf extract of *Jatropha curcas* L. (Euphorbiaceae). International Journal of Applied Research in Natural Products 1(4), 1–4.

Essayan DM, 2001. "Cyclic nucleotide phosphodiesterases". J Allergy Clin Immunol. 108, 671–80.

Etuk EU, Ugwah MO, Ajagbonna OP, Onyeyili PA, 2009. Ethnobotanical survey and preliminary evaluation of medicinal plants with antidiarrhoea properties in Sokoto state, Nigeria. Journal of Medicinal Plants Research 3(10), 763–766.

Evani de LA, Karina PR, José GS, Mendonça RMP, Xavier HS, 2008. Revisão *Acanthospermum hispidum* DC (Asteraceae): perspectives for a phytotherapeutic product. Brazilian Journal of Pharmacy 18, 777–784.

Ezuruike UF, Prieto JM, 2014. The use of plants in the traditional management of diabetes in Nigeria: Pharmacological and toxicological considerations. Journal of Ethnopharmacology 155(2) 855–924.

Fagbohun TR, Odufunwa KT, 2010. Hypoglycemic effect of methanolic extract of *Anacardium occidentale* leaves in alloxan-induced diabetic rats. Nigerian Journal of Physiological Science 25, 87–90.

Fagundes DJ, Taha MO, 2004. Modelo animal de doença: critérios de escolha e espécies de animais de uso corrente. Acta Cirurgica Brasileira *19*(1), 59–65.

Fahey J, 2005. *Moringa oleifera*: A Review of the medical evidence for its nutritional, therapeutic and prophylactic properties. Part 1. Trees for Life Journal 1–15.

Fajemiroye JO *et al.*, 2012. Involvement of 5-HT1A in the anxiolytic-like effect of dichloromethane fraction of Pimenta pseudocaryophyllus. J Ethnopharmacol 3, 872–877.

Fajemiroye JO *et al.*, 2016. Treatment of anxiety and depression: Medicinal plants in retrospect. Fundam Clin Pharmacol doi: 10.1111/fcp.12186.

Fajemiroye JO *et al.*, 2014. Anxiolytic and antidepressant like effects of natural food flavour (E)-methyl isoeugenol. Food Funct. 5, 1819-28.

Fang DC, Hu GX, Hou SX, Hu Y, Jiang MX, 1987. Hemodynamic effects of berberine on conscious rats. Yao Xue Xue Bao = Acta Pharmaceutica Sinica 22(5), 321–5.

Fang DH, Yao WX, Xia GJ, Qu L, Jiang MX, 1981. Studies on the calcium antagonistic action of tetrandrine: I. Effects of tetrandrine on the isolated guinea pig atrium. *Acta Academiae Medicinae Wuhan = Wu-Han I Hsu/eh Yu/an Hsu/eh Pao 1*(2), 46–50.

Faria A, Monteiro R, Azevedo I, Calhau C, 2007a. Pomegranate juice effects on cytochrome P450S expression: *in vivo* studies. J. Med. Food 10, 643–649.

Faria A, Monteiro R, Mateus N, Azevedo I, Calhau C, 2007b. Effect of pomegranate (*Punica granatum*) juice intake on hepatic oxidative stress. Eur. J. Nutr. 46, 271–278.

Farida T, Salawu OA, Tijani AY, Ejiofor JI, 2012. Pharmacological evaluation of *Ipomoea asarifolia* (Desr.) against carbon tetrachloride-induced hepatotoxicity in rats. Journal of Ethnopharmacology 142(3), 642–646.

Farnsworth NR, The role of medicinal plants in drug development. 1984. Natural products and drug development. Proceeding of the Alfred Benzon symposium 20 held at the premises of the Royal Danish Academy of Sciences and Letters. Edited by Krogsgaard-Larsen; Brogger Christensen, P; Kofoed, S; Munksgaard, H., Copenhagen. 1984.

Florentino IF, Silva DPB, Galdino PM, Lino RC, Martins JLR, Silva DM, Costa EA, 2016. Antinociceptive and anti-inflammatory effects of Memora nodosa and allantoin in mice. *Journal of Ethnopharmacology 186*, 298–304.

Foster BC, Foster MS, Vandenhoek S, Krantis A, Budzinski JW, Arnason JT, *et al.*, 2001. An *in vitro* evaluation of human cytochrome P450 3A4 and P-glycoprotein inhibition by garlic. J. Pharm. Pharm. Sci. 4, 176–184.

French ED *et al.*, 1996. Effects of Ibogaine, and cocaine and morphine after Ibogaine, on ventral tegmental dopamine neurons. Life Sci 59: 199–205.

Fuhr U, Kummert AL, 1995. The fate of naringin in humans: a key to grapefruit juice-drug interactions? Clin Pharmacol Ther. 58:365–373.

Fuhr U, Maier A, Blume H, Muck W, Unger S, Staib AH, 1994. Grapefruit juice increases oral nimodipine bioavailability (Abstract) Eur J Clin Pharmacol. 47:100.

Furst PT, 1974. Archeological Evidence for Snuffing in Prehispanic Mexico. In Botanical Museum Leaflets, Harvard University, Vol. 24, pp. 1-28.

Gabriel EI, Chidiebere OF, Ottah A, 2013. Evaluation of the methanolic rhizome extract of *Anchomanes difformis* for analgesic and antipyretic activities. International Journal of Basic and Applied Sciences 2(August 2009), 289–296.

Gad SC *et al.*, 2006. Nonclinical vehicle use in studies by multiple routes in multiple species. Int J Toxicol 25:499–521.

Galey FD, Holstege DM, Plumlee KH, Tor E, Johnson B, Anderson ML, Brown F, 1996. Diagnosis of oleander poisoning in livestock. *Journal of Veterinary Diagnostic Investigation/: Official Publication of the American Association of Veterinary Laboratory Diagnosticians, Inc 8*(3), 358–64.

Galien C. De alimentorum facultatibus, Lib. I, cap. 34, De simplicium medicamentorum temperamentis ac facultatibus, Lib VII, cap. 10 edit. Kuhn, Lipsiae, 1826.

Ganju L, Karan D, Chanda S, Srivastava KK, Sawhney RC, Selvamurthy W, 2003. Immunomodulatory effects of agents of plant origin. Biomed-Pharmacother 57, 296–300.

Garba A, Maurice NA, Maina VA, Baraya YS, Owada AH, Sa'adatu I, Hambolu SE, 2013. Effect of *Leptadenia hastata* leaf extract on embyo-foetal development in white albino rats. Greener Journal of Medical Sciences 3(4), 120–128.

Garrido G, González D, Lemus Y, García D, Lodeiro L, Quintero G, Delgado R, 2004. *In vivo* and *in vitro* anti-inflammatory activity of *Mangifera indica* L. extract (vimang). Pharmacological Research 50(2), 143–149.

Gebreyohannes G, Gebreyohannes M, 2013. Medicinal values of garlic: A review. International Journal of Medicine and Medical Sciences 5(9), 401–408.

Gemedo-Dalle T, Maass BL, Isselstein J, 2005. Plant biodiversity and ethnobotany of Borana pastoralists in southern Oromia, Ethiopia. Economic Botany 59(1), 43–65.

Germano MP, de Pasquale R, Iauk L, Galati EM, Keita A, Sanogo R, 1996. Antiulcer activity of *Vernonia kotschyana* Sch. Bip. Phytomedicine 2(3), 229–233.

Ghate NB, Hazra B, Sarkar R, Mandal N, 2014. Heartwood extract of *Acacia catechu* induces apoptosis in human breast carcinoma by altering bax/bcl-2 ratio. Pharmacognosy Magazine 10(37), 27.

Gheorghiade M, Abraham WT, Albert, NM, Greenberg BH, O'Connor CM, She L, 2006. OPTIMIZE-HF Investigators and Coordinators. Systolic blood pressure at admission, clinical characteristics, and outcomes in patients hospitalized with acute heart failure. *JAMA* 296(18), 2217–26.

Gibson GG, and Skett P, Introductionto Drug Metabolism. London: Nelson Thornes. 2001.

Gilani AH, Aftab K, Suria A, Siddiqui S, Salem R, Siddiqui BS, Faizi S, 1994. Pharmacological studies on hypotensive and spasmolytic activities of pure compounds from *Moringa oleifera*. Phytotherapy Research 8(2), 87–91.

Gilani AH, Shaheen F, Zaman M, Janbaz KH, Shah BH, Akhtar MS, 1999. Studies on antihypertensive and antispasmodic activities of methanol extract of *Acacia nilotica* pods. Phytotherapy Research 13(8), 665–669.

Gills LS, *Ethnomedical uses of plants in Nigeria*. Benin City: Uniben Press. 1992.

Gilman *et al.*, Goodman and Gilman's The Pharmacological Basis of Therapeutics. New York: Pergamon Press. 1992.

Glennon RA *et al.*, 1984. Evidence for 5-HT2 involvement in the mechanism of action of hallucinogenic agents. Life Sciences 35, 2505-2511.

Glesinger L. Medicine through centuries. Zagreb: Zora; 1954. pp. 21–38.

Gomathi D, Ravikumar G, Kalaiselvi M, Devaki K, Uma C, 2013. Efficacy of *Evolvulus alsinoides* (L.) L. on insulin and antioxidants activity in pancreas of streptozotocin induced diabetic rats. Journal of Diabetes and Metabolic Disorders 12(1), 39.

Gondwe M, Kamadyaapaa DR, Tuftsa M, Chuturgoonb AA, Musabayane CT, 2008. *Sclerocarya birrea* (A. Rich.) Hochst. (Anacardiaceae) stem-bark ethanolic extract (SBE) modulates blood glucose, glomerular ltration rate (GFR) and mean arterial blood pressure (MAP) of STZ-induced diabetic rats. Phytomedicine 15, 699–709.

Gouwakinnou GN, Lykke AM, Assogbadjo AE, Sinsin B, 2011. Local knowledge, pattern and diversity of use of *Sclerocarya birrea*. Journal of Ethnobiology and Ethnomedicine 7(1), 8.

Gowri S, Chinnaswamy P, 2011. Evaluation of invitro antimutagenic activity of *Caralluma adscendens* Roxb. in bacterial reverse mutation assay. Journal of Natural Product and Plant Resources 1(4), 27–34.

Goyal BR, Mahajan SG, Mali RG, Goyal RK, Mehta AA, 2007. Global Journal of Pharmacology 1(1), 6-12.

Graber-Maier A, Buter KB, Aeschlimann J, Bittel C, Kreuter M, Drewe J, *et al.*, 2010. Effects of Curcuma extracts and curcuminoids on expression of P-glycoprotein and cytochrome P450 3A4 in the intestinal cell culture model LS180. Planta Med. 76, 1866–1870.

Graybiel AM, 2008. Habits, rituals, and the evaluative brain. Annual Review of Neuroscience 31, 359–387.

Greeske K, Pohlmann BK, 1996. Horse chestnut seed extract—an effective therapy principle in general practice. Drug therapy of chronic venous insufficiency. *Fortschritte Der Medizin 114*(15), 196–200.

Griebel G and Holmes A, 2013. 50 years of hurdles and hope in anxiolytic drug discovery. Nature Reviews Drug Discovery 12, 667–687.

Grob P, Häcki M, Müller-Schoop J, Joller-Jemelka H, 1975. Drug-induced pseudolupus. *The Lancet 306*(7926), 144–148.

Grosvenor PW, Supriono A, Gray DO, 1995. Medicinal plants from Riau Province, Sumatra, Indonesia. Part 2: Antibacterial and antifungal activity. Journal of Ethnopharmacology 45(2), 97–111.

Guarim NGO, 2006. Saber tradicional pantaneiro: as plantas medicinais e a educação ambiental. Revista Eletrônica do Mestrado em Educação Ambiental 17.

Guevara AP, Vargas C, Sakurai H, Fujiwara Y, Hashimoto K, Maoka T, Nishino H, 1999. An antitumor promoter from *Moringa oleifera* Lam. Toxicology and Environmental Mutagenesis 440(2), 181–188.

Guleria S, Tiku AK, Singh G, Vyas D, Bhardwaj A, 2011. Antioxidant activity and protective effect against plasmid DNA strand scission of leaf, bark, and heartwood extracts from *Acacia catechu*. Journal of Food Science 76(7), 959–964.

Gupta A, Gautam MK, Singh RK, Kumar MV, Rao CV, Goel RK, Anupurba S, Gunjam M, Ravindran M, Sengamalam R, Goutam, KJ, Jha AK, 2010. Pharmacognostic and antidiabetic study of Clitoria ternatea. International Journal of Phytomedicine 2: 373-378.

Gupta R, Mathur M, Bajaj VK, Katariya P, Yadav S, Kamal R, Gupta RS, 2012. Evaluation of antidiabetic and antioxidant activity of *Moringa oleifera* in experimental diabetes. Journal of Diabetes 4(2), 164–171.

Haidara K, Zamir L, Shi QW, Batist G, 2006. The flavonoid Casticin has multiple mechanisms of tumor cytotoxicity action. Cancer Lett. 242, 180–190.

Hajda J, Rentsch KM, Gubler C, Steinert H, Stieger B, Fattinger K, 2010. Garlic extract induces intestinal P-glycoprotein, but exhibits no effect on intestinal and hepatic CYP3A4 in humans. Eur. J. Pharm. Sci. 41, 729–735.

Hamburger and Hostettmann. 1991. Bioactivity in Plants: The link between phytochemistry and medicine. Phytochemistry 30 (12): 3864-3874.

Handa S. S., Khanuja S. P. S., Longo G., Rakesh D. D., 2008. Extraction technologies for medicinal and aromatic plants, ICS-UNIDO, Trieste

Han GQ, Wei LH, Li CL, Qiao L, Jia YZ, Zheng QT, 1989. The isolation and identification of PAF inhibitors from Piper wallichii (Miq.) Hand-Mazz and P. hancei Maxim. *Yao Xue Xue Bao = Acta Pharmaceutica Sinica 24*(6), 438–43.

Hanley MJ, Masse G, Harmatz JS, Court MH, Greenblatt DJ, 2012. Pomegranate juice and pomegranate extract do not impair oral clearance of flurbiprofen in human volunteers: divergence from *in vitro* results. Clin. Pharmacol. Ther. 92, 651–657.

Hari Kumar KB, Kuttan R, 2006. Inhibition of drug metabolizing enzymes (cytochrome P450) *in vitro* as well as *in vivo* by Phyllanthus amarus SCHUM and THONN. Biol. Pharm. Bull. 29, 1310–1313 10.1248/bpb.29.1310

Harouna S, Adama H, Pierre S, Eric A, Moussa C, Germaine NO, 2012. Dyeing and medicinal plants used in the area of Mouhoun in Burkina Faso. Universal Journal of Environmental Research and Technology 2(3), 110–118.

Hasrat JA *et al.*, 1997. Isoquinoline derivatives isolated from the fruit of Annona muricata as 5-HTergic 5-HT1A receptor agonists in rats: unexploited antidepressive (lead) products. J Pharm Pharmacol 49, 1145–1149.

Hazra B, Sarkar R, Biswas S, Mandal N, 2010. The antioxidant, iron chelating and DNA protective properties of 70 per cent methanolic extract of 'Katha' (Heartwood extract of *Acacia catechu*). Journal of Complementary and Integrative Medicine 7(1).

He SM, Li CG, Liu JP, Chan E, Duan, W, and Zhou SF, 2010. Disposition pathways and pharmacokinetics of herbal medicines inhumans. Curr. Med. Chem. 17, 4072–4113.

He W, Van Puyvelde L, Bosselaers J, De Kimpe N, Van der Flaas M, Roymans A, Chalo Mutiso PB, 2002. Activity of 6-pentadecylsalicylic acid from *Ozoroa insignis* against marine crustaceans. Pharmaceutical Biology 40(1), 74–76.

Hebbani AV, Shridhar S, Jahnavi PB, Nallanchakravarthula V, 2014. Protective effect of *Evolvulus alsinoides* (Linn) and Decalepis hamiltoni Wight and Arn aqueous extracts against hydrogen peroxide induced oxidative damage of human erythrocytes. International Journal of Research in BioScience 3(2), 18–28.

Hegazy SK, El-Bedewy M, Yagi A, 2012. Antifibrotic effect of aloe vera in viral infection-induced hepatic periportal fibrosis. World J. Gastroenterol. 18, 2026–2034.

Heikal L, Starr A, Martin GP, Nandi M, Dailey LA, 2016. *In vivo* pharmacological activity and biodistribution of S-Nitrosophytochelatins after intravenous and intranasal administration in mice. *Nitric Oxide/: Biology and Chemistry/Official Journal of the Nitric Oxide Society.*

Hellberg K, Ruschewski W, de Vivie R, 1975. Drug induced acute renal failure after heart surgery (author's transl). *Thoraxchirurgie, Vaskula/re Chirurgie* 23(4), 396–9.

Herekrishna R, Anup C, Satyabrata B, Bhabani SN, Sruti RM, Ellaiah P, 2010. Preliminary phytochemical investigation and anthelmintic activity of *Acanthospermum hispidum* DC. Journal of Pharmaceutical Science and Technology 2(5), 217–221.

Hernández EX, Exploración Etnobotánica y su Metodología, Colegio de Posgraduados, Escuela Nacional de Agricultura, Chapingo, México. 1970.

Hewageegana HGS, Ariyawansa HAS, Ratnasooriya WD, 2006. Gastroprotective activity of the paste of *Evolvulus alsinoides* L. Vidyoka Journal of Science 13, 23–31.

Hickling K and Smith D, The contribution of vehicles, rates of administration, and volumes to infusion studies. In: Healing G, Smith D, editors. Handbook of preclinical intravenous infusion. New York (NY): Taylor and Francis. 2000.

Higuchi T and Connors KA, 1965. Phase-solubility techniques. Adv Anal Chem Inst 4, 117–212.

Ho BE, Shen DD, McCune JS, Bui T, Risler L, Yang Z, *et al.*, 2010. Effects of garlic on cytochromes P450 2C9- and 3A4-mediated drug metabolism in human hepatocytes. Sci. Pharm. 78, 473–481.

Hogan MC, Foreman KJ, Naghavi M, Ahn SY, Wang M, Makela SM, Murray CJL, 2010. Maternal mortality for 181 countries, 1980-2008: a systematic analysis of progress towards Millennium Development Goal 5. Lancet 375(9726), 1609–23.

Hu, Z, Yang X, Ho PC, Chan SY, Heng PW, Chan E, *et al.*, 2005. Herb-drug interactions: a literature review. Drugs 65, 1239–1282.

Huang L, Liu JF, Liu LX, Li DF, Zhang Y, Nui HZ, Zhang CY, 1993. Antipyretic and anti-inflammatory effects of Artemisia annua L. *China Journal of Chinese Materia Medica* 18(1), 44–48.

Huei-Chen H, Chai-Rong L, Pei-Dawn, PDLC, Ching-Chow C, Shu-Hsun C, 1991. Vasorelaxant effect of emodin, an anthraquinone from a Chinese herb. *European Journal of Pharmacology* 205(3), 289–294.

Huerkamp MJ, 2002. Alcohol as a disinfectant for aseptic surgery of rodents: crossing the thin blue line? Contemp Top Lab Anim Sci 41:10–12.

Huie CW, 2002. A review of modern sample-preparation techniques for the extraction and analysis of medicinal plants. Anal Bioanal Chem. 373, 23-30.

Hussaini HSN, Karatela YY, 1989. Traditional medicinal plants used by Hausa tribe of Kano state of Nigeria. Interational Journal of Drug Research 27(4), 211–216.

Hussein G, Miyashiro H, Nakamura N, Hattori M, Kakiuchi N, Shimotohno K, 2000. Inhibitory effects of Sudanese medicinal plant extracts on hepatitis C virus (HCV) protease. Phytotherapy Research 14(7), 510–516.

Hussein G, Miyashiro H, Nakamura N, Hattori M, Kawahata T, Otake T, Shimotohno K, 1999. Inhibitory effects of Sudanese plant extracts on HIV1 replication and HIV1 protease. Phytotherapy Research 13(1), 31–36.

Ibrahim G, Abdurahman HI, Ibrahim NDG, Yaro H, 2009. Studies on Acute Toxicity and Anti-Inflammatory Effects of *Vernonia kotschyana* Sch. Bip. (Asteraceae). Nigerian Journal of Pharmaceutical Science 8(2), 8–12.

Idohou-Dossou N, Diouf A, Gueye A, Guiro A, Wade S, 2011. Impact of daily consumption of Moringa (*Moringa oleifera*) dry leaf powder on iron status of Senegalese lactating women. African Journal of Food, Agriculture, Nutrition and Development 11(4), 4985–4999.

Idu M, Erhabor JO, Efijuemue HM, 2010. Documentation on medicinal plants sold in markets in Abeokuta, Nigeria. Tropical Journal of Pharmaceutical Research 9(2), 110–118.

Igbinosa OO, Igbinosa IH, Chigor VN, Uzunuigbe OE, Oyedemi SO, Odjadjare EE, Igbinosa EO, 2011. Polyphenolic contents and antioxidant potential of stem bark extracts from *Jatropha curcas* (Linn). International Journal of Molecular Sciences 12(5), 2958–2971.

Igbinosa OO, Oviasogie EF, Igbinosa EO, Igene O, Igbinosa IH, Idemudia OG, 2013. Effects of Biochemical Alteration in Animal Model after Short-Term Exposure of *Jatropha curcas* (Linn) Leaf Extract. The Scientific World Jornal 2013, 1–5.

Ignacimuthu S, Ayyanar M, Sankarasivaraman K, 2008. Ethnobotanical study of medicinal plants used by Paliyar tribals in Theni district of Tamil Nadu, India. Fitoterapia 79(7), 562–568.

Igoli JO, Ogaji OG, Tor-Anyiin TA, Igoli NP, 2005. Traditional medicine practice amongst the Igede people of Nigeria. Part II. African Journal of Traditional, Complementary and Alternative Medicines 2(2), 134–152.

Ikhiri K, Boureima D, Dicko D, Koulodo D, 1992. Chemical screening of medicinal plants used in the traditional pharmacopoeia of Niger. Pharmaceutical Biology 30(4), 251–262.

ILAR, Guide for the care and use of laboratory animals, chapter 2: animal care and use program. Washington (DC): National Academies Press. 2010.

Indhumol VG, Pradeep HR, Sushrutha CK, Jyothi T, Shavas MM, 2013. Ethnomedicinal, Phytochemical, and Therapeutic applications of. International Research Journal of Pharmacy and Plant Science 1(2), 1–6.

Inngjerdingen K, Nergård CS, Diallo D, Mounkoro PP, Paulsen BS, 2004. An ethnopharmacological survey of plants used for wound healing in Dogonland, Mali, West Africa. Journal of Ethnopharmacology 92(2-3), 233–44.

Inngjerdingen KT, Meskini S, Austarheim I, Ballo N, Inngjerdingen M, Michaelsen TE, Paulsen BS, 2012. Chemical and biological characterization of polysaccharides from wild and cultivated roots of *Vernonia kotschyana*. Journal of Ethnopharmacology 139(2), 350–358.

Innocent E, Hassanali A, Kisinza WNW, Mutalemwa PPP, Magesa S, Kayombo E, 2014. Anti-mosquito plants as an alternative or incremental method for malaria vector control among rural communities of Bagamoyo District, Tanzania. Journal of Ethnobiology and Ethnomedicine 10(1), 56.

Ioannidis JPA, 2005. Why most published research findings are false. PLoS Med 2: 696–701.

Ionescu C, and Caira MR, Drug Metabolism: Current Concepts. Dordrecht: Springer. 2005.

Ip C, Lisk DJ, 1997. Modulation of phase I and phase II xenobiotic-metabolizing enzymes by selenium-enriched garlic in rats. Nutr. Cancer 28, 184–188.

Irwin S, 1962. Drug screening and evaluative procedures. Science 136, 123-28.

Irwin S, 1968. Comprehensive observational assessment: Ia. A systematic, quantitative procedure for assessing the behavioral and physiologic state of the mouse. Psychopharmacologia (Berl) 13, 222-57.

Ishak CY, Mohammed S, Ayoub H, 2013. Antimicrobial activity of four medicinal plants used by Sudanese traditional medicine. Journal of Forest Products and Industries 2(1), 29–33.

Iskandar I, Hadju V, As S, Natsir R, 2015. Effect of *Moringa oleifera* leaf extracts supplementation in preventing maternal anemia. International Journal of Scientific and Research Publications 5(2), 5–7.

Islam MR, Mannan MA, Kabir MHB, Islam A, Olival KJ, 2010. Analgesic, anti-inflammatory and antimicrobial effects of ethanol extracts of mango leaves. Journal of the Bangladesh Agricultural University 8(2), 239– 244.

Iwu MM, 1993. *Handbook of African Medicinal Plants*. Florida, USA: CRC Press.

Iyer, D., Patil, U.K. 2011. Effect of *Evolvulus alsinoides* L. ethanolic extract and its fraction in experimentally induced hyperlipidemia in rats. Pharmacology Online 580, 573–580.

Jaffe AM, Gephardt D, Courtemanche L, 1990. Poisoning due to ingestion of Veratrum viride (false hellebore). *The Journal of Emergency Medicine 8*(2), 161–7.

Jain A, Katewa SS, Galav P, Nag A, 2008. Some therapeutic uses of biodiversity among the tribals of Rajasthan. Indian Journal of Traditional Knowledge 7(2), 256–262.

Jain SK, Srivastava S, 2005. Traditional use of some Indian plants among islanders of the Indian Ocean. Indian Journal of Traditional Knowledge 4, 345–357.

Jain SK, Dictionary of Indian folk medicine and ethnobotany. Deep Publications, New Delhi, India. 1991.

Jaiswal D, Rai PK, Kumar A, Mehta S, Watal G, 2009. Effect of *Moringa oleifera* Lam. leaves aqueous extract therapy on hyperglycemic rats. Journal of Ethnopharmacology 123(3), 392–396.

Jayakumar T, Sridhar MP, Bharathprasad TR, Ilayaraja M, Govindasamy S, Balasubramanian MP, 2009. Journal of Health Science 55(5), 701-708.

Jayakumar G, Ajithabai MD, Sreedevi S, Viswanathan PK, Remeshkumar B, 2010. Ethnobotanical survey of the plants used in the treatment of diabetes of. Indian Journal of Traditional Knowledge 9(January), 100–104.

Jayanthi MK, Jyoti, MB, 2012. Experimental animal studies on analgesic and anti-nociceptive activity of Allium sativum (Garlic) powder. Indian Journal of Rresearch in Reproductive Medical Science 2(1), 1–6.

Jegede IA, Nwinyi FC, Ibrahim J, Ugbabe G, Dzarma S, Kunle OF, 2009. Investigation of phytochemical, anti inflammatory and anti nociceptive properties of *Ipomoea asarifolia* leaves. Journal of Medicinal Plant Research 3(3), 160–165.

Jeréz Bezzenberger J, Plantas Mágicas de la Costa Valdiviana, Ed. Kultrún, Valdivia, Chile. 2005.

Jiangsu New Medical College. *A compendium of Chinese medicine* (ed). Hong Kong: Shanghai Scientific and Technological I Press; 1979.

Jigam AA, Akanya HO, Dauda BE, Okogun JO, 2010. Polygalloyltannin isolated from the roots of *Acacia nilotica* Del.(Leguminoseae) is effective against *Plasmodium berghei* in mice. Journal of Medicinal Plant Research 4(12), 1169–1175.

Jin MW, Zong XG, Fang DC, Jiang MX, 1985. Effect of rhomotoxin on the transmembrane potential and contractile force of guinea pig papillary muscles. *Yao Xue Xue Bao = Acta Pharmaceutica Sinica* 20(7), 481–4.

Jiofack T, Fokunang C, Guedje N, Kemeuze V, Fongnzossie E, Nkongmeneck BA, Tsabang N, 2010. Ethnobotanical uses of medicinal plants of two ethnoecological regions of Cameroon. Journal of Medicine and Medical Sciences 2(March), 60–79.

John D, 1984. One hundred useful raw drugs of the Kani tribes of Trivandrum Forest Division, Kerala, India. Int. J. Crude Drug Res. 22(1), 17–39.

Joona K, Sowmia C, Kp D, Mj D, 2013. Preliminary phytochemical investigation of *Mangifera indica* leaves and screening of antioxidant and anticancer activity. Research Journal of Pharmaceutical, Biological and Chemical Sciences 4(1), 1112–1118.

Josefsson M, Zackrisson AL, Ahlner J, 1996. Effect of grapefruit juice on the pharmacokinetics of amlodipine in healthy volunteers. Eur J Clin Pharmacol. 51:189–193.

Joseph G, Zhao Y, Klaus W, 1995. Pharmacologic action profile of crataegus extract in comparison to epinephrine, amirinone, milrinone and digoxin in the isolated perfused guinea pig heart. *Arzneimittel-Forschung* 45(12), 1261–5.

Jussofie A, *et al.*, 1994. Kavapyrone enriched extract from Piper methysticum as modulator of the GABA binding site in different regions of rat brain. Psychopharmacology (Berl) 116, 469–474.

Juteau F, Masotti V, Bessiere JM, Dherbomez M, Viano J, 2002. Antibacterial and antioxidant activities of *Artemisia annua* essential oil. Journal of Ethnopharmacology 73, 532–535.

Juyal P, Ghildiyal JC, 2013. Medicinal phyto-diversity of Bhabar tract of Garhwal Himalaya. Journal of Medicinal Plants Studies 1(6), 43–57.

Kaithwas G, Dubey K, Pillai KK, 2011. Effect of aloe vera (Aloe barbadensis Miller) gel on doxorubicin-induced myocardial oxidative stress and calcium overload in albino rats. Indian J. Exp. Biol. 49, 260–268.

Kamagate M, Bamba-kamagate D, Die-kacou H, Ake-assi L, Yavo J, Daubret-potey T, Haramburu F, 2015. Pharmacovigilance of medicinal plants. International Journal of Phytopharmacology 6(2), 66–75.

Kamanzi Atindehou K, Schmid C, Brun R, Koné MW, Traore D, 2004. Antitrypanosomal and antiplasmodial activity of medicinal plants from Côte d'Ivoire. Journal of Ethnopharmacology 90(2-3), 221–227.

Kamba AS, Hassan LG, 2010. Phytochemical screening and antimicrobial activities of *Euphorbia balsamifera* leaves, stems and root against some pathogenic microorganisms. African Journal of Pharmacy and Pharmacology 4(9), 645–652.

Kambizi L, Afolayan AJ, 2001. An ethnobotanical study of plants used for the treatment of sexually transmitted diseases in Guruve District, Zimbabwe. Journal of Ethnopharmacology 77, 5–9.

Kamuhabwa A, Nshimo C, De Witte P, 2000. Cytotoxicity of some medicinal plant extracts used in Tanzanian traditional medicine. Journal of Ethnopharmacology 70(2), 143–149.

Kankara SS, Ibrahim MH, Mustafa M, Go R, 2015. Ethnobotanical survey of medicinal plants used for traditional maternal healthcare in Katsina state, Nigeria. South African Journal of Botany 97, 165–175.

Karou SD, Tchacondo T, Ilboudo DP, Simpore J, 2011. Sub Saharan Rubiaceae: A review of their traditional uses, phytochemistry and biological activities. Pakistan Journal of Biological Sciences 12(3), 149–169.

Katz RJ, 1981. Animal models and human depressive disorders. Neurosci Biobehav Rev 5, 231–246.

Kaur K, Michael H, Arora S, Härkönen P, Kumar S, 2005. In vitro bioactivity-guided fractionation and characterization of polyphenolic inhibitory fractions from *Acacia nilotica* (L.) Willd. ex Del. Journal of Ethnopharmacology 99(3), 353–360.

Kaur S, Mondal P, 2014. Study of total phenolic and flavonoid content, antioxidant activity and antimicrobial properties of medicinal plants. Journal of Microbiology and Experimentation 1(1), 1–6.

Kawashima K, Hayakawa T, Miwa Y, Oohata H, Suzuki T, Fujimoto K, Chen ZX, 1990. Structure and hypotensive activity relationships of tetrandrine derivatives in stroke-prone spontaneously hypertensive rats. *General Pharmacology 21(3)*, 343–7.

Kecskeméti V, Braquet P, 1990. Electrophysiological effects of platelet-activating factor (PAF) and its antagonist, BN 52021 in cardiac preparations. *European Journal of Pharmacology 183(4)*, 1240–1241.

Khamis S, *et al.*, 2004. Phytochemistry and preliminary biological evaluation of Cyathostemma argenteum, a Malaysian plant used traditionally for the treatment of breast cancer. Phytother Res 18, 507–510.

Khan M, Khan A, Najeeb-ur-Rehman, Gilani A-H, 2015. Blood pressure lowering, vasodilator and cardiac-modulatory potential of Carum roxburghianum seed extract. *Clinical and Experimental Hypertension (New York, N.Y./: 1993) 37(2)*, 102–7.

Khan MA, Khan T, Ahmad Z, 1994. Barks used as source of medicine in Madhya Pradesh, India. Fitoterapia 65, 444– 446.

Khosravi-Boroujeni H, Sarrafzadegan N, Mohammadifard N, Sajjadi F, Maghroun M, Asgari S, Azadbakht L, 2013. White rice consumption and CVD risk factors among Iranian population. *Journal of Health, Population, and Nutrition 31(2)*, 252–61.

Kilkenny C, Browne W, Cuthill IC, Emerson M and Altman DG, Animal research: Reporting in vivo experiments: The ARRIVE guidelines. British Journal of Pharmacology (2010) 160, 1577–1579.

Kimura Y, Ito H, Hatano T, 2010. Effects of mace and nutmeg on human cytochrome P450 3A4 and 2C9 activity. Biol. Pharm. Bull. 33, 1977–1982.

King VF, Garcia ML, Himmel D, Reuben JP, Lam YK, Pan JX, Kaczorowski GJ, 1988. Interaction of tetrandrine with slowly inactivating calcium channels. Characterization of calcium channel modulation by an alkaloid of Chinese medicinal herb origin. The Journal of Biological Chemistry 263(5), 2238–44.

Kisangau DP, Lyaruu HVM, Hosea KM, Joseph CC, 2007. Use of traditional medicines in the management of HIV/AIDS opportunistic infections in Tanzania: a case in the Bukoba rural district. Journal of Ethnobiology and Ethnomedicine 3, 29.

Kleijnen J, Knipschild P, 1992. Ginkgo biloba. *The Lancet* 340(8828), 1136–1139.

Kleijnen J, Knipschild P, ter Riet G, 1989. Garlic, onions and cardiovascular risk factors. A review of the evidence from human experiments with emphasis on commercially available preparations. *British Journal of Clinical Pharmacology* 28(5), 535–44.

Klohs MW, *et al.*, 1959. A chemical and pharmacological investigation of Piper methysticum Forst. J Med Pharm Chem 1, 95–103.

Ko FN, Lin CN, Liou SS, Huang TF, Teng CM, 1991. Vasorelaxation of rat thoracic aorta caused by norathyriol isolated from Gentianaceae. *European Journal of Pharmacology* 192(1), 133–9.

Koltai M, *et al.*, 1991. Platelet-activating factor (PAF): a review of its effects, antagonists and future clinical implications. Drugs 29, 174- 204.

Komada M, *et al.*, 2008. Elevated Plus Maze for Mice. J Vis Exp. 2008 (22): 1088.

Konan NA, Bacchi EM, 2007. Antiulcerogenic effect and acute toxicity of a hydroethanolic extract from the cashew (*Anacardium occidentale* L.) leaves. Journal of Ethnopharmacology 112(2), 237–242.

Konan NA, Bacchi EM, Lincopan N, Varela SD, Varanda EA, 2007. Acute, subacute toxicity and genotoxic effect of a hydroethanolic extract of the cashew (*Anacardium occidentale* L.). Journal of Ethnopharmacology 110(1), 30–38.

Koné WM, Kamanzi Atindehou K, 2008. Ethnobotanical inventory of medicinal plants used in traditional veterinary medicine in Northern Côte d'Ivoire (West Africa). South African Journal of Botany 74(1), 76–84.

Koné WM, Kamanzi Atindehou K, Terreaux C, Hostettmann K, Traoré D, Dosso M, 2004. Traditional medicine in North Côte-d'Ivoire: Screening of 50 medicinal plants for antibacterial activity. Journal of Ethnopharmacology 93(1), 43–49.

Kostewicz ES, *et al.*, 2002. Forecasting the oral absorption behavior of poorly soluble weak bases using solubility and dissolution studies in biorelevant media. Pharm Res 19, 345–349.

Koukouikila-Koussounda F, Abenab AA, Nzounganic A, Mombouli JV, Ouambae JM, Kunf J, Ntoumia F, 2013. *In vitro* evaluation of antiplasmodial activity of extracts of *Acanthospermum hispidum* DC [Asteraceae] and *Ficus thonningii* Blume [Moraceae], two plants used in traditional medicine in the republic of Congo. African Journal of Traditional, Complementary and Alternative Medicines 10(2), 270–276.

Kudi AC, Myint SH, 1999. Antiviral activity of some Nigerian medicinal plant extracts. Journal of Ethnopharmacology 68(1-3), 289–294.

Kuehl GE, Lampe JW, Potter JD, and Bigler J, 2005. Glucuronidation of nonsteroidal anti-inflammatory drugs: identifying the enzymes responsible in human liver microsomes. Drug Metab. Dispos. 33, 1027–1035.

Kulkarni AS, Mute VM, Dhamane SP, Gadekar AS, 2012. Evaluation of antibacterial activity of *Caralluma adscendens* Roxb. Stem. International Research Journal of Pharmacy 3(8), 209–211.

Kumar Y, Haridasan K, Rao RR, 1980. Ethnobotanical notes on certain medicinal plants among some Garo people around Balphakram Sanctuary in Meghalaya. Bull. Bot. Surv. India 22, 161–165.

Kumar PS, Mishra D, Ghosh G, Panda CS, 2010. Medicinal uses and pharmacological properties of *Moringa oleifera*. International Journal of Phytomedicine 2(3).

Kuo C, Lee C, Lin S, *et al.*, 2010. Rapid determination of aristolochic acids I and II in herbal products and biological samples by ultra-high-pressure liquid chromatography-tandem mass spectrometry. Talanta. 80:1672e1680.

Kuramochi T, Chu J, Suga T, 1994. Gou-teng (from Uncaria rhynchophylla Miquel)-induced endothelium-dependent and -independent relaxations in the isolated rat aorta. *Life Sciences 54*(26), 2061–9.

Kuribara H, Iwata H, Tomioka H, Takahashi R, Goto K, Murohashi N, Koya S, 2001. Effects of antiinflammatory triterpenes isolated from *Leptadenia hastata* latex on keratinocyte proliferation. Phytotherapy Research 15(2), 131–134.

Kwaji A, Japaru M, 2015. Phytochemical analysis and antimicrobial activity of *Albizia chevalieri* (Harms) stem bark methanol extract. Topclass Journal of Herbal Medicine 4(2), 16–19.

Kwan CY, Daniel EE, Chen, MC, 1990. Inhibition of Vasoconstriction by Tetramethylpyrazine. *Journal of Cardiovascular Pharmacology 15*(1), 157–162.

Lai L, Lin L, Lin J, Tsai T, 2003. Pharmacokinetic study of free mangiferin in rats by microdialysis coupled with microbore high-performance liquid chromatography and tandem mass spectrometry. J Chromatogr A. 987:367e374.

Lakshmi DU, Adilaxmamma A, Reddy AG, Rao VV, 2011. Evaluation of herbal methionine and *Mangifera indica* against lead-induced organ toxicity in broilers. Toxicology International 18(1), 58–61.

Lakshmi TL, Aravind kumar S, 2011. Preliminary phytochemical analysis and *in vitro* antibacterial activity of *Acacia catechu* willd bark against *Streptococcus mitis, Streptococcus sanguis* and *Lactobacillus acidophilus*. International Journal of Phytomedicine 3(4), 579–584.

Lamxay V, de Boer JH, 2011. Traditions and plant use during pregnancy, childbirth and postpartum recovery by the Kry ethnic group in Lao PDR. Journal of Ethnobiology and Ethnomedicine 7:14.

Langford SD, Boor PJ, 1996. Oleander toxicity: an examination of human and animal toxic exposures. *Toxicology 109*(1), 1–13.

Latha S, Rajaram K, Suresh KP, 2014. Hepatoprotective and antidiabetic effect of methanol extract of *Caralluma*. International Journal of Pharmacy and Pharmaceutical Sciences 6(1), 10–13.

Le Bon AM, Vernevaut MF, Guenot L, Kahane R, Auger J, Arnault I, *et al.*, 2003. Effects of garlic powders with varying alliin contents on hepatic drug metabolizing enzymes in rats. J. Agric. Food Chem. 51, 7617–7623.

Lee KJ, You HJ, Park SJ, Kim YS, Chung YC, Jeong TC, *et al.*, 2001. Hepatoprotective effects of Platycodon grandiflorumon acetaminophen-Induced liver damage in mice. Cancer Lett. 174, 73–81.

Lekshmi UMD, Reddy PN, 2011. *Evolvulus alsinoides'un* antiinflamatuar, antipiretik ve antidiyare. Turkish Journal of Biology 35(5), 611–618.

Lemonica IP, Alvarenga CMD, 1994. Abortive and teratogenic effect of *Acanthospermum hispidum* DC. and *Cajanus cajan* (L.) Millps. in pregnant rats. Journal of Ethnopharmacology 43(1), 39–44.

Liang YZ, Xie P, Chan K 2004; Quality control of herbal medicines J Chromatogr B, 812: 53–70.

Li CX, Zhao GS, Li XG, 1987. Effects of sinomenine on action potential and force of contraction in guinea pig heart muscle. *Yao Xue Xue Bao = Acta Pharmaceutica Sinica 22*(8), 561–5.

Li CY, Devappa RK, Liu JX, Lv JM, Makkar HPS, Becker K, 2010. Toxicity of *Jatropha curcas* phorbol esters in mice. Food and Chemical Toxicology 48(2), 620–625.

Li GR, Qian JQ, Lü FH, 1990. Effects of neferine on heart electromechanical activity in anaesthetized cats. *Zhongguo Yao Li Xue Bao = Acta Pharmacologica Sinica 11*(2), 158–61.

Li H, Zhang BH, 1989. The antiarrhythmic effect of sophoramine. *Yao Xue Xue Bao = Acta Pharmaceutica Sinica* 24(2), 147–50.

Li JX *et al.*, 2005. Five new oleanolic acid glycosides from Achyranthes bidentata with inhibitory activity on osteoclast formation. Planta Med. 71, 673–9.

Li P, *et al.*, 1999. Evaluation of intravenous flavopiridol formulations. PDA J Pharm Sci Technol 53, 137–140.

Li P and Zhao L, 2007. Developing early formulations: practice and perspective. Int J Pharm 341, 1–19.

Li Y, Zhang Y, Wang Z, Zhu J, Tian Y, Chen B, 2012. Quantitative analysis of atractylenolide I in rat plasma by LC-MS/MS method and its application to pharmacokinetic study. J Pharm Biomed Anal. 58:172e176.

Li YJ, Deng HW, Chen X, 1987. The protective effect of ginsenosides and its components on myocytes anoxia/reoxygenation and myocardial reperfusion injury. *Yao Xue Xue Bao = Acta Pharmaceutica Sinica* 22(1), 1–5.

Lifongo LL, Simoben CV, Ntie-Kang F, Babiaka SB, Judson PN, 2014. A bioactivity versus ethnobotanical survey of medicinal plants from Nigeria, West Africa. Natural Products and Bioprospecting 4, 1–19.

Lipinski CA *et al.*, 2001. Experimental and computational approaches to estimate solubility and permeability in drug discovery and development settings. Adv Drug Deliv Rev 46, 3–26.

Lipp F, 2002. Herbalism, Taschen GmbH, Germany. 2002.

Lister RG, 1987. The use of a plus-maze to measure anxiety in the mouse. Psychopharmacology 92, 180–185.

Litovitz TL, Clark LR, Soloway RA, 1993. Annual Report of the American Association of Poison Control Centers Toxic Exposure Surveillance System. *Am J Emerg Med.* 12546– 584.

Liu D, Zhao GS, 1987. Effects of dl-tetrahydropalmatine on cardiac hemodynamics in dogs]. *Yao Xue Xue Bao = Acta Pharmaceutica Sinica* 22(7), 537–40.

Liu JQ, Yang YF, Wang CF, Li Y, Qiu MH, 2012. Three new diterpenes from *Jatropha curcas*. Tetrahedron 68(4), 972–976.

Liu KCSC, Yang SL, Roberts MF, Elford BC, Phillipson JD, 1992. Antimalarial activity of *Artemisia annua* flavonoids from whole plants and cell cultures. *Plant Cell Reports* 11(12), 637–640.

Liu SF, Cai YN, Evans TW, McCormack DG, Barer GR, Barnes PJ, 1990. Ligustrazine is a vasodilator of human pulmonary and bronchial arteries. *European Journal of Pharmacology* 191(3), 345–50.

Liu XC, Li YP, Li HQ, Deng ZW, Zhou L, Liu ZL, Du SS, 2013. Identification of repellent and insecticidal constituents of the essential oil of *Artemisia rupestris* L. aerial parts against liposcelis bostrychophila Badonnel. *Molecules 18*(9), 10733–10746. http://doi.org/10.3390/molecules180910733

Lizardi-Ramos C, 1969. El mural de los bebedores en Cholula, Puebla. En Boletín Bibliográfico de la SHCP, año XV, 2ª. Época, núm. 426, pp. 14-15.

Loh SH, Lee AR, Huang WH, Lin CI, 1992. Ionic mechanisms responsible for the antiarrhythmic action of dehydroevodiamine in guinea-pig isolated cardiomyocytes. *British Journal of Pharmacology* 106(3), 517–23.

Lompo-Ouedraogo Z, van der Heide D, van der Beek EM, Swarts HJM, Mattheij JAM, Sawadogo L, 2004. Effect of aqueous extract of *Acacia nilotica* ssp adansonii on milk production and prolactin release in the rat. The Journal of Endocrinology 182(2), 257–266.

Londhe VP, Gavasane AT, Nipate SS, Bandawane DD, Chaudhari PD, 2011. Review role of garlic (*Allium sativum*) in various diseases/: an overview. Journal of Pharmaceutical Research and Opinion 4, 129–134.

Lown KS, Bailey DG, Fontana RJ, *et al.*, 1997. Grapefruit juice increases felodipine oral availability in humans by decreasing intestinal CYP3A protein expression. J Clin Invest. 99:2545–2553.

Lozoya X, 2003. Las Plantas del Alma, Arqueología Mexicana, Vol. X, No. 59, enero-febrero, INAH, México, pp. 58-63.

Lozoya X, 1999. Un Paraíso de Plantas Medicinales, Arqueología Mexicana, Vol. VII, No. 39, septiembre-octubre, INAH, México, 14-21.

Lu YM, Liu GQ, 1990. The effects of (–)-daurisoline on Ca2+ influx in presynaptic nerve terminals. *British Journal of Pharmacology* 101(1), 45–48.

Lucki I, 2001. A prescription to resist proscription for murine models of depression. Psychopharmacology 153, 395–398.

Lundahl J, Regardh CG, Edgar B, Johnsson G, 1997. Effects of grapefruit juice ingestion–pharmacokinetics and haemodynamics of intravenously and orally administered felodipine in healthy men. Eur J Clin Pharmacol. 52:139–145.

Lundahl J, Regardh CG, Edgar B, Johnsson G, 1995. Relationship between time of intake of grapefruit juice and its effect on pharmacokinetics and pharmacodynamics of felodipine in healthy subjects. Eur J Clin Pharmacol. 49:61–67.

Maden K, Plant Collection and Herbarium Techniques. Our Nature. 2004, 2, 53-57.

Madge C, 1998. Therapeutic landscapes of the Jola, The Gambia, West Africa. Health and Place 4(4), 293–311.

Madhavarao B, Sabithadevi K, Vinnakoti 2011. In –vitro antimicrobial and free radical scavenger assay of two medicinal plants Clitoria ternatea and Cardiospermum halicacabum, International Journal of Chemical and Analytical Science 2(11): 1253-1255.

Magassouba FB, Diallo A, Kouyaté M, Mara F, Mara O, Bangoura O, Baldé AM, 2007. Ethnobotanical survey and antibacterial activity of some plants used in Guinean traditional medicine. Journal of Ethnopharmacology 114(1), 44–53.

Maheshu V, Priyadarsini DT, Sasikumar JM, 2014. Antioxidant capacity and amino acid analysis of *Caralluma adscendens* (Roxb.) Haw var. fimbriata (wall.) Grav. and Mayur. aerial parts. Journal of Food Science and Technology 51(10), 2415–2424.

Mahmoudvand H, Sepahvand P, Jahanbakhsh S, Azadpour M, 2014. Evaluation of the antileishmanial and cytotoxic effects of various extracts of garlic (*Allium sativum*) on *Leishmania tropica*. Journal of Parasitic Diseases 5, 20–29.

Maitra I, *et al.*, 1995. Peroxyl radical scavenging activity of ginkgo biloba extract EGb761. Biochem Pharmacol. 49, 1649-1655.

Makare N, Bodhankar S, Rangari V, 2001. Immunomodulatory activity of alcoholic extract of Mangifera indica L. in mice. Journal of Ethnopharmacology 78(2-3), 133–137.

Malone MH and Robichaud RC, 1983. The pharmacological evaluation of natura products - General and specific approaches to screening ethnopharmaceuticals. J. Ethnopharmacol, 8, 127-147.

Malviya S, Rawat S, Kharia A, Verma M, 2011. Medicinal properties of *Acacia nilotica* Linn.- A comprehensive review on ethnopharmacological claims. International Journal of Pharmacy and Life Sciences 2(6), 830–837.

Manisha S, Rashiya B, Manoj M, 2012. A Case study of medicinal plants used by local women for gynecological disorders in Karaikal (U.T. of Puducherry). Journal of Phytology 4(5): 09-12.

Manjrekar AP, Jisha V, Bag PP, Adhikary B, Pai MM, Hegde A, *et al.*, 2008. Effect of Phyllanthus niruri Linn. treatment on liver, kidney and testes in CCl4 induced hepatotoxic rats. Indian J. Exp. Biol. 46, 514–520.

Marcondes DBS, Reichert CL, Andrade LF, Santos CAM, Weffort-Santos AM, 2014. Cytotoxicity and apoptogenic effectsof *Lafoensia pacari*. Journal of Ethnopharmacology 157, 243-250.

Mariani Ramírez C, Temas de hipnosis. Andrés Bello (Universidad Católica), Santiago, Chile. 1965.

Markowitz JS, Devane CL, Chavin KD, Taylor RM, Ruan Y, Donovan JL, 2003. Effects of garlic (*Allium sativum* L.) supplementation on cytochrome P450 2D6 and 3A4 activity in healthy volunteers. Clin. Pharmacol. Ther. 74, 170–177.

Martindale, The Extra Pharmacopoeia, The Pharmaceutical Press, London.

Marques LJ *et al.*, 1999. "Pentoxifylline inhibits TNF-alpha production from human alveolar macrophages". Am. J. Respir. Crit. Care Med. 159, 508–11.

Masamha B, Gadzirayi CT, Mukutirwa I, 2010. Efficacy of *Allium sativum* (garlic) in controlling nematode parasites in sheep. International Journal of Applied Research in Veterinary Medicine 8(3), 161–169.

Masferrer Kan E, 2003. Los Alucinógenos en las Culturas Contemporáneas, Arqueología Mexicana, Vol. X, No. 59, enero-febrero, INAH, México, pp. 50-55.

Masferrer Kan E, and Díaz Brenis E, 2000. De lo sagrado a lo perverso. La manipulación de las tradiciones indígenas por Occidente. En R. Gutiérrez y M. Villalobos, Espiritualidad de los pueblos indígenas de América, Universidad Michoacana de San Nicolás de Hidalgo, Morelia, México, pp. 167-171.

Mash DC, Kovera CA, Pablo J, Tyndale RF, Ervin FD, and Williams IC, 2000. Ibogaine: Complex Pharmacokinetics, concerns for safety and preliminary efficacy measures. Ann NY Acad Sci, 914, 394-401.

Massadeh AM, Al-Safi SA, Momani IF, Alomary AA, Jaradat QM, Alkofahi AS, 2007. Garlic (*Allium sativum* L.) as a potential antidote for cadmium and lead intoxication: Cadmium and lead distribution and analysis in different mice organs. Biological Trace Element Research 120(1-3), 227–234.

Mashour NH, Lin GI, Frishman WH, Astin JA, Ernest E, Tyler VE, Jones TK, 1998. Herbal Medicine for the Treatment of Cardiovascular Disease. Archives of Internal Medicine 158(20), 2225.

Massiha A, Khoshkholgh-pahlaviani MM, Issazadeh K, Bidarigh S, Zarrabi S, 2013. Antibacterial activity of essential oils and plant extracts of Artemisia (*Artemisia annua* L.) *in vitro*. Zahedan Journal of Research in Medical Sciences 15(6), 14–18.

Mathieu G, Meissa D, 2011. Traditional leafy vegetables in Senegal: diversity and medicinal uses. African Journal of Traditional, Complementary and Alternative Medicines 4(August), 1–7.

Mavundza EJ, Maharaj R, Finnie JF, Kabera G, Van Staden J, 2011. An ethnobotanical survey of mosquito repellent plants in Mkhanyakude district, KwaZulu-Natal province, South Africa. Journal of Ethnopharmacology 137(3), 1516–1520.

Maxia A, Lancioni MC, Balia AN, Alborghetti R, Pieroni A, Loi MC, 2008. Medical ethnobotany of the Tabarkins, a Northern Italian (Ligurian) minority in south-western Sardinia. Genetic Resources and Crop Evolution 55(6), 911–924.

Mazzari AL, and Prieto JM, 2014. Herbal medicines in Brazil: pharmacokinetic profile and potential herb–drug interactions. Front. Pharmacol. 5, 162.

McCarthy AM, Femoral cannulation using the tail cuff exteriorization method in the rat, p 20–25. In: Healing G, Smith D, editors. Handbook of preclinical continuous intravenous infusion. New York (NY): Taylor and Francis; 2000.

McGaw LJ, Eloff JN, 2008. Ethnoveterinary use of southern African plants and scientific evaluation of their medicinal properties. Journal of Ethnopharmacology 119(3), 559–574.

Medeiros RMT, Barbosa RC, Riet-Correa F, Lima EF, Tabosa IM, De Barros SS, Molyneux RJ, 2003. Tremorgenic syndrome in goats caused by *Ipomoea asarifolia* in Northeastern Brazil. Toxicon 41(7), 933–935.

Mehta P, Shah R, Lohidasan S, Mahadik KR, 2015. Pharmacokinetic profile of phytoconstituent(s) isolated from medicinal plants: A comprehensive review. J. Tradit. Complement. Med. 5:207-227.

Meier zu Biesen C. 2011. The rise to prominence of *Artemisia annua* L. – the transformation of a Chinese plant to a global pharmaceutical. African Sociological Review/Revue Africaine de Sociologie 14(2), 24–46.

Meira M, da Silva EP, David JM, David JP, 2012. Review of the genus Ipomoea: Traditional uses, chemistry and biological activities. Brazilian Journal of Pharmacognosy 22(3), 682–713.

Melillo De Magalhães P, Dupont I, Hendrickx A, Joly A, Raas T, Dessy S, Schneider YJ, 2012. Anti-inflammatory effect and modulation of cytochrome P450 activities by Artemisia annua tea infusions in human intestinal Caco-2 cells. Food Chemistry 134(2), 864–871.

Mendonça EAF, Coelho MFB, Luchese M, 2006. Teste de tetrazólio em sementes de mangaba-brava (Lafoensia pacari St. Hil. - Lythraceae). Revista Brasileira de Plantas Medicinais 8(2), 33-38.

Mercier AP, Los Secretos de los Chamanes Mayas, Ediciones Luciérnaga, España, 2009. pp. 54-56.

Miccadei S, Di Venere D, Cardinali A, Romano F, Durazzo A, Foddai MS, *et al.*, 2008. Antioxidative and apoptotic properties of polyphenolic extracts from edible part of artichoke (*Cynara scolymus* L.) on cultured rat hepatocytes and on human hepatoma cells. Nutr. Cancer 60, 276–283.

Middelkoop TB and Labadie RP, 1985. The action of Saraca asoca Roxb. de Wilde bark on the PGH2 synthetase enzyme complex of the sheep vesicular gland. Z. Naturforsch. C 40, 523–526.

Mikail HG, 2010. Phytochemical screening, elemental analysis and acute toxicity of aqueous extract of *Allium sativum* L. bulbs in experimental rabbits. Journal of Medicinal Plants Research 4(4), 322–326.

Mirdeilami SZ, Barani H, Mazandarani M, Heshmati GA, 2011. Ethnopharmacological Survey of Medicinal Plants in Maraveh Tappeh Region, North of Iran. Iranian Journal of Plant Physiology 2(1), 327–338.

Mishra SB, Vijaykumar M, Ojha SK, Verma A, 2013. Antidiabetic effect of *Jatropha* curcas L. leaves extract in normal and Alloxan induced diabetic rats. International Journal of Pharmaceutical Scienes 5(2), 2081–2085.

Modarai M, Suter A, Kortenkamp A, Heinrich M. 2011. The interaction potential of herbal medicinal products: a luminescence-based screening platform assessing effects on cytochrome P450 and its use with devil's claw (Harpagophyti radix) preparations. J. Pharm. Pharmacol. 63, 429–438.

Modern Pharmacognosy, E. Ramstad, McGraw-Hill, New York.

Mohamed IET, El Nur EBES, Abdelrahman MEN, 2010. The antibacterial, antiviral activities and phytochemical screening of some Sudanese medicinal plants. EurAsian Journal of BioSciences 16(4), 8–16.

Mohan S, Thiagarajan K, Chandrasekaran R, Arul J, 2014. In vitro protection of biological macromolecules against oxidative stress and in vivo toxicity evaluation of *Acacia nilotica* (L.) and ethyl gallate in rats. BMC Complementary and Alternative Medicine 14(1), 257.

Mølgaard P, Nielsen SB, Rasmussen DE, Drummond RB, Makaza N, Andreassen J, 2001. Anthelmintic screening of Zimbabwean plants traditionally used against schistosomiasis. Journal of Ethnopharmacology 74(3), 257– 264.

Mollik MAH, Hossan MS, Paul AK, Taufiq-Ur-Rahman, Jahan R. and Rahmatullah M, 2010. A comparative analysis of medicinal plants used by folk medicinal healers in three districts of Bangladesh and inquiry as to mode of selection of medicinal plants. Ethnobotany Res. Appl. 8, 195–218. 11.

Molyneaux, BL, The Sacred Earth, Evergreen. 1995, pp. 144-145.

Montecino S, Mitos de Chile. Diccionario de seres, magia y encantos, Ed. Sudamericana, Santiago, Chile. 2003.

Montúfar López A, 1998. Arqueobotánica del centro ceremonial de Tenochtitlán, Arqueología Mexicana, Vol. VI, No. 31, mayo-junio, pp. 34-41.

Montúfar López A, 1985. Estudio de los restos vegetales recolectados en la cueva de las ventanas, Chihuahua. En A. Montúfar (coord.) Estudios palinológicos y paleoetnobotánicos, INAH, México, pp. 111-133.

Morris R, 1984. Developments of a water-maze procedure for studying spatial learning in the rat. J Neurosci Methods 11, 47-60.

Morton DB. A systematic approach for establishing humane endpoints. ILAR J. 2000 41:80–86.

Morton JF, 1991. The horseradish tree, Moringa pterygosperma (Moringaceae)–a boon to arid lands? Economic Botany 45(3), 318–333.

Mösbach EW, Botánica Indígena de Chile, Museo Chileno de Arte Precolombino, Fundación Andes, Andrés Bello, Santiago, Chile. 1999.

Moshi MJ, Cosam JC, Mbwambo ZH, Kapingu M, Nkunya MH, 2004. Testing beyond ethnomedical claims: brine shrimp lethality of some Tanzanian plants. Pharmaceutical Biology 42(7), 547–551.

Moshi MJ, Otieno DF, Weisheit A, 2012. Ethnomedicine of the Kagera Region, north western Tanzania. Part 3: plants used in traditional medicine in Kikuku village, Muleba District. Journal of Ethnobiology and Ethnomedicine 8(1), 14.

Mouren X, Caillard P, Schwartz F, 1994. Study of the antiischemic action of EGb 761 in the treatment of peripheral arterial occlusive disease by TcPo2 determination. *Angiology 45(6), 413–7.*

Mowrey DB, *Herbal Tonic Therapies- Remedies from nature's own pharmacy to strengthen and amp; support each vital body system.* Connecticut: Keats Publishing. 1993.

Moxley RA, Schneider NR, Steinegger DH, Carlson MP, 1989. Apparent toxicosis associated with lily-of-the-valley (Convallaria majalis) ingestion in a dog. *Journal of the American Veterinary Medical Association 195(4), 485–7.*

Mugisha MK, Asiimwe S, Namutebi A, Borg-Karlson AK, Kakudidi EK, 2014. Ethnobotanical study of indigenous knowledge on medicinal and nutritious plants used to manage opportunistic infections associated with HIV/AIDS in western Uganda. Journal of Ethnopharmacology 155(1), 194–202.

Müller F, 1972. Estudios iconográficos del mural de Los Bebedores, Cholula, Puebla, en Religión en Mesoamérica. XII Mesa Redonda, t. 1, Sociedad Mexicana de Antropología, México, pp. 141-146.

Muller V, Chávez JH, Reginatto FH, Zucolotto SM, Niero R, Navarro D, Yunes RA, Schenkel EP, Barardi CRM, Zanetti CR, Simões CMO, 2007. Evaluation of antiviral activity of South American plant extracts against herpes simplex virus type 1 and rabies virus. Phytotherapy Research 21, 970-974.

Multidisciplinary Association for Psychedelic Research (MAPS). From <http://www.maps.org.> (Retrieved on 14 June 2016).

Musa A, Ibrahim M, Aliyu A, Abdullahi M, Tajuddeen N, Ibrahim H, Oyewale A, 2015. Chemical composition and antimicrobial activity of hexane leaf extract of *Anisopus mannii* (Asclepiadaceae). Journal of Intercultural Ethnopharmacology 4(2), 1.

Musa MS, Abdelrasool FE, Elsheikh EA, Ahmed LAMN, Mahmoud ALE, Yagi SM, 2011. Ethnobotanical study of medicinal plants in the Blue Nile State, Southeastern Sudan. Journal of Medicinal Plant Research 5(17), 4287–4297.

Musa AM, Aliyu AB, Yaro AH, Magaji MG, Hassan HS, Abdullahi MI, 2009. Preliminary phytochemical, analgesic and anti- inflammatory studies of the methanol extract of *Anisopus mannii* (N.E. Br.) (Asclepiadaceae) in rodents. African Journal of Pharmacy and Pharmacology 3(8), 374–378.

Museo de Arte Precolombino and Ministerio de Educación de Chile. Mapuche: Seeds of The Chilean Soul. An exhibit at the Port of History Museum, Philadelphia, Pennsylvania, USA. 1986.

Mustapha AA, 2014. Ethnobotanical field survey of medicinal plants used by traditional medicine practitioners to manage HIV/AIDS opportunistic infections and their prophylaxis in Keffi Metropolis, Nigeria. Asian Journal of Plant Science and Research 4(1), 7–14.

Muthee JK, Gakuya DW, Mbaria JM, Kareru PG, Mulei CM, Njonge FK, 2011. Ethnobotanical study of anthelmintic and other medicinal plants traditionally used in Loitoktok district of Kenya. *Journal of Ethnopharmacology* 135(1), 15–21.

Muthu C, Ayyanar M, Raja N, Ignacimuthu S, 2006. Medicinal plants used by traditional healers in Kancheepuram district of Tamil Nadu, India. Journal of Ethnobiology and Ethnomedicine 2, 43.

Nadembega P, Boussim JI, Nikiema JB, Poli F, Antognoni F, 2011. Medicinal plants in Baskoure, Kourittenga Province, Burkina Faso: an ethnobotanical study. Journal of Ethnopharmacology 133(2), 378–95.

Naganuma M, Saruwatari A, Okamura S, Tamura H, 2006. Turmeric and curcumin modulate the conjugation of 1-naphthol in Caco-2 cells. Biol. Pharm. Bull. 29, 1476–1479.

Nahata A, Patil UK, Dixit VK, 2009. Anxiolytic activity of *Evolvulus alsinoides* and *Convulvulus pluricaulis* in rodents. Pharmaceutical Biology 47(5), 444–451.

Naidu NR, Bhat S, Dsouza U, 2013. Evaluation of possible protective and therapeutic influence of *Evolvulus alsinoides* on learning and memory against Alzheimer's disease induced y Aluminium choloride in rat. International Journal of Pharmaceutical and, 1(2), 7–20.

Namuli A, Abdullah N, Sieo CA, Zuhainis SA, Oskoueian E, 2011. Phytochemical compounds and antibacterial activity of *Jatropha curcas* Linn. extracts. Journal of Medicinal Plants Research 5(June 2015), 3982–3990.

Nasa Y, Hashizume H, Hoque AN, Abiko Y, 1993. Protective effect of crataegus extract on the cardiac mechanical dysfunction in isolated perfused working rat heart. *Arzneimittel-Forschung* 43(9), 945–9.

Nath M, Choudhury MD, 2010. Ethno-medico-botanical aspects of Hmar tribe of Cachar district, Assam (Part 1). Indian Journal of Traditional Knowledge 9, 760–764.

Nayak BS, Patel KN, Nayak BS, Trust V, 2010. Anti-inflammatory screening of *Jatropha curcas* root, stem and leaf in albino rats. Romania Journal of Biology–Plant Biology 55, 9–13.

Ndamba J, Nyazema N, Makaza N, Anderson C, Kaondera KC, 1994. Traditional herbal remedies used for the treatment of urinary schistosomiasis in Zimbabwe. Journal of Ethnopharmacology 42(2), 125–132.

Nda-Umar UI, Gbate M, Umar AN, Mann A, 2014. Ethnobotanical study of medicinal plants used for the treatment of malaria in Nupeland, north central Nigeria. Global Journal of Research on Medicinal Plants and Indigenous Medicine 3(4), 112–126.

Neafsey P, Ginsberg G, Hattis D, Johns DO, Guyton KZ, and Sonawane B, 2009. Genetic polymorphism in CYP2E1: Population distribution of CYP2E1 activity. J. Toxicol. Environ. Health B Crit. Rev. 12, 362–388.

Nedelcheva A, Draganov S, 2014. Ethnobotany and Biocultural Diversities in the Balkans. In *Bulgarian Medical ethnobotany: The Power of plants in Pragmatic and Poetic Frames*, pp. 45–65.

Neervannan S, 2006. Preclinical formulations for discovery and toxicology: physicochemical challenges. Expert Opin Drug Metab Toxicol 2, 715–731.

Nergard CS, Diallo D, Michaelsen TE, Malterud KE, Kiyohara H, Matsumoto T, Paulsen BS, 2004. Isolation, partial characterisation and immunomodulating activities of polysaccharides from *Vernonia kotschyana* Sch. Bip. ex Walp. Journal of Ethnopharmacology 91(1), 141–152.

Neuwinger HD, *African Ethnobotany: Poisons and Drugs: Chemistry, Pharmacology, Toxicology*. New York: Chapman and Hall. 1996.

Neuwinger HD, *African traditional medicine. A dictionary of plant use and application.* Stuttgart: Medpharm GmbH Scientific Publishers. 2000.

Ni Y, *et al.*, 1996. Preventive effect of Ginkgo biloba extract on apoptosis in rat cerebellar neuronal cells induced by hydroxyl radicals. Neurosci Lett. 214, 115-118.

Nie SQ, Xie ZC, Lin KC, 1985. Effects of tetrapyrazine on membrane fluidity and electrophoretic mobility of platelets and the relation to its antiaggregation effect. *Yao Xue Xue Bao = Acta Pharmaceutica Sinica* 20(9), 689–92.

NIH. Guidelines for the Use of Non-Pharmaceutical-Grade Chemicals/Compounds in Laboratory Animals. Animal Research Advisory Committee, Office of Animal Care and Use, NIH; 2008.

NIH. National Institute of Drug Abuse. Retrieved on 20 June 2016 from www: <https://www.drugabuse.gov/publications/drugfacts/hallucinogens>.

Nikam, P H, Kareparamban, J, Jachav, A and Kadam, V 2012: Future Trends in Standardisation of Herbal Drugs. *J App Pharm Sci.* 02 (06) 38-44

Nishioka S. de A, Resende ES, 1992. Transitory complete atrioventricular block associated to ingestion of Nerium oleander. *Revista Da Associac'/aƀo Meìdica Brasileira (1992)* 41(1), 60–2.

Niu Z, Chen F, Sun J, *et al.*, 2008. High-performance liquid chromatography for the determination of 3-n-butylphthalide in rat plasma by tandem quadrupole mass spectrometry: application to a pharmacokinetic study. J Chromatogr B. 870:135e139.

Nogueira NP, Reis PA, Laranja GA, Pinto AC, Aiub CA, Felzenszwalb I, *et al.*, 2011. *In vitro* and *in vivo* toxicological evaluation of extract and fractions from Baccharis trimera with anti-inflammatory activity. J. Ethnopharmacol. 138, 513–522.

NRC. Occupational Health and Safety in the Care and Use of Research Animals. Washington: National Academy Press; 1997.

NRC. Occupational Health and Safety in the Care and Use of Nonhuman Primates. Washington: National Academies Press; 2003.

Nutrients GE, 2014. Notice to US Food and Drug Administration that the use of Slimaluma, a hydroethanolic extract of *Caralluma fimbriata*, is generally recognised safe.

Nwauzoma AB, Dappa MS, 2013. Ethnobotanical Studies of Port Harcourt Metropolis, Nigeria. ISRN Botany 2013, 1–11.

Nwodo N, Ibezim A, Ntie-Kang F, Adikwu M, Mbah C, 2015. Anti-trypanosomal activity of Nigerian plants and their constituents. Molecules 20(5), 7750–7771.

Oates J, Hardman J, Limbird L, Molinoff P, Antihypertensive agents and the drug therapy of hypertension. In Goodman and Gilman's (Ed.), *The Pharmacological Basis of Therapeutics* (9th ed., pp. 781–808). New York, NY: McGraw-Hill Book Co. 1996.

Obata OO, Aigbokhan EI, 2012. Ethnobotanical practices among the people of Okaakoko, Nigeria. Plant Archives 12(2), 627–638.

Ocvirk S, Kistler M, Khan S, Talukder SH, Hauner H, 2013. Traditional medicinal plants used for the treatment of diabetes in rural and urban areas of Dhaka, Bangladesh – an ethnobotanical survey. Journal of Ethnobiology and Ethnomedicine 9(43), 1–8.

Odugbemi TO, Odunayo RA, Ibukun EA, Fabeku PO, 2007. Medicinal plants useful for malaria therapy in Okeigbo, Ondo state, southwest Nigeria. African Journal of Traditional Complementary Alternative Medicine 4(2), 191–198.

Ody P, *The Complete Medicinal Herbal (1993). Dorling Kindersley.* New York; 1993.

Ogbole OO, Adebayo AG, Ajaiyeoba EO, 2010. Ethnobotanical survey of plants used in treatment of inflammatory diseases in Ogun State of Nigeria. European Journal of Scientific Research 43(2), 183–191.

Ogie-Odia EA, Oluowo EF, 2009. Assessment of some therapeutic plants of the Abbi people in Ndokwa West L.G.A of Delta State, Nigeria. Ethnobotanical Leaflets 13, 989–1002.

Ogwang PE, Ogwal JO, Kasasa S, Olila D, Ejobi F, Kabasa D, Obua C, 2012. *Artemisia annua* L. infusion consumed once a week reduces risk of multiple episodes of malaria: A randomised trial in a Ugandan community. Tropical Journal of Pharmaceutical Research 11(3), 445–453.

Ojewole JAO, 2006. Evaluation of the anticonvulsant effect of Sclerocarya birrea (A. Rich.) [Anacardiaceae] stem-bark aqueous extract in mice. Natural Medicine 61, 67–72.

Ojowele JAO, Mawoza T, Chiwororo WDH, Owira PMO, 2010. *Sclerocarya birrea* (A. Rich) Hochst. ["Marula"] (Anacardiaceae): A review of its phytochemistry, pharmacology and toxicology and its ethnomedicinal uses. Phytotherapy Research ;24(5): 633-9.

Okechukwu PCU, Nzubechukwu E, Ogbanshi ME, 2015. The effect of ethanol leaf extract of Jatropha curcas on chloroform induced hepatotoxicity in Albino rats. Global Journal of Biotechnology and Biochemistry 10(1), 11–15.

Okon U, Etim B, 2014. *Citrus aurantifolia* impairs fertility facilitators and indices in male albino Wistar rats. International Journal of Reproduction, Contraception, Obstetrics and Gynecology 3(3), 640–645.

Okpanachi AO, Adelaiye AB, Dikko AAU, Kabiru M, Mohammed A, Tanko Y, 2010. Evaluation of the Effect of aqueous-methanolic stem bark extract of *Acacia polyacantha* on blood glucose levels of alloxan induced diabetic Wistar rats. International Journal of Animal and Animal Veterinary Advances 2(3), 59–62.

Okpo SO, Ching FP, Ayinde BA, Udi OO, Alonge PO, Eze GO, 2011. Gastroprotective effects of the ethyl acetate fraction of *Anchomanes difformis* (Engl). International Journal of Health Research 4(4), 155–161.

Olajide OB, Fadimu OY, Osaguona PO, Saliman MI, 2013. Ethnobotanical and phytochemical studies of some selected species of Leguminoseae of Northern Nigeria: a study of Borgu Local Government Area, Niger State. Nigeria. International Journal of Science and Nature 4(3), 546–551.

Olatokun WM, Ayanbode O, 2009. Use of indigenous knowledge by women in a Nigerian rural community. Indian Journal of Traditional Knowledge 8(2), 287–295.

Olivares JC, El Umbral Roto. Escritos en Antropología Poética. Fondo Matta – Museo Chileno de Arte Precolombino, Santiago, Chile. 1995.

Olowokudejo JD, Kadiri AB, Travih VA, 2008. An ethnobotanical survey of herbal markets and medicinal plants in Lagos State of Nigeria. Ethnobotanical Leaflets 12, 851–865.

Oluranti AC, Michael UO, Jane UC, 2012. Ethnobotanical studies of medicinal plants used in the management of Peptic ulcer disease in Sokoto State, North Western Nigeria. International Research Journal of Pharmacy and Pharmacology 2(9), 225–230.

Oluwole OG, Esume C, 2015. Anti-inflammatory effects of aqueous extract of Mangifera indica in Wistar rats. Journal of Basic and Clinical Physiology and Pharmacology 26(3), 313– 315.

Ong ES, 2004. Extraction methods and chemical standardization of botanicals and herbal preparations. J Chromatogr B Analyt Technol Biomed Life Sci. 812, 23-33.

Ong HC, Norzalina J, 1999. Malay herbal medicine in Gemencheh, Negri Sembilan, Malaysia. Fitoterapia 70(1), 10–14.

Ortiz de Montellano BR, Aztec Medicine, Health, and Nutrition, Rutgers Universty Press, New Brunswick, New Jersey. 1990.

Oskoueian E, Abdullah N, Saad WZ, Omar AR, Kuan WB, Zolkifli NA, … Ho YW, 2011. Antioxidant, anti-inflammatory and anticancer activities of methanolic extracts from *Jatropha curcas* Linn. Journal of Medicinal Plants Research 5(1), 49–57.

Osman HM, Shayoub ME, Babiker EM, Osman B, Elhassan AM, 2012. Effect of ethanolic leaf extract of *Moringa oleifera* on aluminum-induced anemia in white albino rats. Jordan Journal of Biological Siences 5(4), 255–260.

Othman AR, Abdullah N, Ahmad S, Ismail IS, Zakaria MP, 2015. Elucidation of in-vitro anti-inflammatory bioactive compounds isolated from *Jatropha curcas* L. plant root. BMC Complementary and Alternative Medicine 15(1), 1–10.

Otieno JN, Kennedy MMH, Herbert VL, Mahunnah RLA, 2011. Multi-plant or single-plant extracts, which is the most effective for local healing in tanzania? African Journal of Traditional, Complementary and Alternative Medicines 5(2), 1–7.

Otitoju O, Nwamarah JU, Otitoju GTO, Okorie AU, Stevens C, and Baiyeri KP, 2014. Effect of *Moringa oleifera* aqueous leaf extract on some haematological indices in Wistar rats. Chemical and Process Engineering Research 18, 26–30.

Ott J, 1996. 'Entheogens II: On entheology and ethnobotany', Journal of Psychoactive Drugs, 28, pp. 205–9.

Ouattara L, Koudou C, Zango C, Barro N, Savadogo A, Bassole IHN, Traore AS, 2011. Antioxidant and antibacterial activities of three species of *Lannae* from Burkina Faso. Journal of Applied Science 11(1), 157–162.

Ouédraogo S, Ranaivo HR, Ndiaye M, Kaboré ZI, Guissou IP, Bucher B, Andriantsitohaina R, 2004. Cardiovascular properties of aqueous extract from *Mitragyna inermis* (Willd). Journal of Ethnopharmacology 93(2-3), 345–350.

Oyama Y *et al.*, 1994. Myricetin and quercetin, the flavonoid constituents of Ginkgo biloba extract, greatly reduce oxidative metabolism in both resting and Ca2+ loaded brain neurons. Brain Res. 635, 125- 129.

Oyewo OO, Onyije FM, Ashamu EA, OWA, Ayeni OJ, 2012. Evaluation of aqueous stem bark extract of *Mangifera indica* in the liver of Wistar rats. Asian Journal of Pharmaceutical and Biological Research 2(4), 231–233.

Ozaki Y, 1990. Vasodilative effects of indole alkaloids obtained from domestic plants, Uncaria rhynchophylla Miq. and Amsonia elliptica Roem. et Schult. *Nihon Yakurigaku Zasshi. Folia Pharmacologica Japonica 95*(2), 47–54.

Pacher P, Bátkai S, and Kunos G. 2006. The Endocannabinoid System as an Emerging Target of Pharmacotherapy, National Institute of Health, Pharmacol Rev, September, 58 3, 389-462.

Palladino L, *Vine of the Soul: A Phenomenological Study of Ayahuasca and its Effect on Depression*. PhD Dissertation. California, USA: Pacifica Graduate Institute. 2009.

Pari L, Karamaæ M, Kosińska A, Rybarczyk A, Amarowicz R, 2007. Antioxidant activity of the crude extracts of drumstick tree (*Moringa oleifera*) Polish Journal of Food and Nutrition Sciences 57(2), 203–208.

Parimaladevi B, Boominathan R, Mandal SC, 2003. Antiinflammatory, analgesic and anti-pyretic properties of Clitoria ternatea root, Fitoterapia 74: 345-349.

Park JH, Park YK, Park E, 2009. Antioxidative and antigenotoxic effects of garlic (*Allium sativum* L.) prepared by different processing methods. Plant Foods for Human Nutrition 64(4), 244–249.

Parveen BU, Shikha RAK, 2007. Traditional uses of medicinal plants among the rural communities of Churu district in the Thar Desert, India. Journal of Ethnopharmacology 113, 387–399.

Pashler H, Wagenmakers EJ, 2012. Editors' introduction to the special section on replicability in psychological science: a crisis of confidence? Perspect Psychol Sci 7, 528–530.

Patel NB, 2000. Mechanism of action of cathinone: the active ingredient of khat (*Cathaedulis*). East Afr Med J 77, 329–332.

Patil P S and Shettigar R 2010: An Advancement of Analytical Techniques in Herbal Research *J. Adv. Sci. Res*, 1(1); 08-14

Paul D, De D, Ali KM, Chatterjee K, Nandi DK, Ghosh D, 2010. Contraception 81(4), 355-361.

Pedersen ME, Vestergaard HT, Hansen SL, Bah S, Diallo D, Jäger AK, 2009. Pharmacological screening of Malian medicinal plants used against epilepsy and convulsions. Journal of Ethnopharmacology 121(3), 472–475.

Perry P, 2007. The ethics of animal research: A UK perspective. ILAR J 48:42-46.

Pert CB, and Snyder SH, 1973. Opiate receptor: demonstration in nervous tissue. Science 179, 1011– 1014.

Peter M, 2013. Ethnobotanical study of some selected medicinal plants used by traditional healers in Limpopo Province (South Africa). American Journal of Research Communication 1(3), 8–23.

Peters-Golden M, *et al.*, 2005. "Leukotrienes: underappreciated mediators of innate immune responses". J Immunol. 174, 589–94.

Pharmacognosy, G.E. Trease and W.C. Evans, Saunders, London.

Pharmacology, D. of. (1979). A clinical study of the antihypertensive effect of tetrandrine. Chinese Medical Journal, 92(3), 193–8.

Pharmacopeial Convention, 1998. Rauwolfia alkaloids USP DI Drug Information for the Health Care Professional. Health Care Professional, 118.

Pi E, Chen J, Huang J, Hou A, 2010. A fast and sensitive HPLC-MS/MS analysis and preliminary pharmacokinetic characterization of cudratricusxanthone B in rats. J Chromatogr B. 878:1953e1958.

Pöpping S, Rose H, Ionescu I, Fischer Y, Kammermeier H, 1995. Effect of a hawthorn extract on contraction and energy turnover of isolated rat cardiomyocytes. *Arzneimittel-Forschung 45*(11), 1157–61.

Porsolt RD, 2000. Animal models of depression: Utility for transgenic research. Rev Neurosci 11, 53–58.

Porsolt RD *et al.*, 1978. "Behaviour Despair in Rats: A New Model Sensitive to Antidepressant Treatments," European Journal of Pharmacology 47, 379-391.

Porsolt RD, *et al.*, 1977a. "Depression: A New Animal Model Sensitive to Antidepressant Treatments," Nature 266, 5604, 730-732.

Porsolt RD, *et al.*, 1977b. "Behaviour Despair in Mice: A Primary Screening Test for Antidepressants," Archives Internationals de Pharmacodynamie et de Therapie 229, 327-336.

Potrich FB, Allemand A, da Silva LM, Dos Santos AC, Baggio CH, Freitas CS, *et al.*, 2010. Antiulcerogenic activity of hydroalcoholic extract of Achillea millefolium L.: involvement of the antioxidant system. J. Ethnopharmacol. 130, 85–92.

Pradesh A, 2011. Ethnobotany of the Monpa ethnic group. Journal of Ethnobiology and Ethnomedicine 31, 1–14.

Pradhan P, Joseph L, Gupta V, Chulet R, Arya H, Verma R, and Bajpai A, 2009. Saraca asoca (Ashoka): a review. J. Chem. Pharm. Res. 1, 62–71.

Prasad S, Kalra N, Singh M, Shukla Y, 2008. Protective effects of lupeol and mango extract against androgen induced oxidative stress in Swiss albino mice. Asian Journal of Andrology 10(2), 313–318.

Prassas I, Diamandis EP, 2008. Novel therapeutic applications of cardiac glycosides. *Nature Reviews Drug Discovery 7*(11), 926–935.

Preethi KC, Kuttan R, 2009. Hepato and reno protective action of Calendula officinalis L. flower extract. Indian J. Exp. Biol. 47, 163–168.

Qin Z, Den H, Zhuang H, 1990. Effect of oxymatrine on prolonging the survival time of cardiac tissue allograft in mice and its immunologic mechanisms. *Zhong Xi Yi Jie He Za Zhi = Chinese Journal of Modern Developments in Traditional Medicine/ Zhongguo Zhong Xi Yi Jie He Yan Jiu Hui (Chou), Zhong Yi Yan Jiu Yuan, Zhu Ban* 10(2), 99–100, 70.

Rabelo AF, Guedes MM, Tome Ada R, Lima PR, Maciel MA, Lira SR, *et al.*, 2010. Vitamin E ameliorates high dose trans-dehydrocrotonin-associated hepatic damage in mice. Nat. Prod. Commun. 5, 523–528.

Radford DJ, Gillies AD, Hinds JA, Duffy P, 1986. Naturally occurring cardiac glycosides. *The Medical Journal of Australia 144*(10), 540–4.

Sathish KR, Rahman AA, Buvanendran R, Obeth D, Panneerselvam U, 2010. Effect of *Evolvulus alsinoides* root extracts on acute reserpine induced orofacial dyskinesia. International Journal of Pharmacy and Pharmaceutical Sciences 2(4), 2–5.

Rajendran S, Deepalakshmi PD, Parasakthy K, Devaraj H, Devaraj SN, 1996. Effect of tincture of Crataegus on the LDL-receptor activity of hepatic plasma membrane of rats fed an atherogenic diet. *Atherosclerosis 123*(1-2), 235–41.

Rajkapoor B, Jayakar B, Murugesh N, Sakthisekaran D, 2006. Chemoprevention and cytotoxic effect of Bauhinia variegata against N-nitrosodiethylamine induced liver tumors and human cancer cell lines. J. Ethnopharmacol. 104, 407–409.

Ramírez E, 2003. El Toloache o Yerba del Diablo, Arqueología Mexicana, Vol. X, No. 59, enero-febrero, INAH, México, pp. 56.

Rao M-R, Shen X.-H, Zou X, 1988. Effects of praeruptorin C and E isolated from "Qian-Hu" on swine coronary artery and guinea-pig atria. European Journal of Pharmacology 155(3), 293–296.

Raponda-Walker A, Sillans R, 1961. Les plantes utiles du Gabon. Quarterly Journal of Crude Drug Research 1(1), 27–27.

Rashid J, McKinstry McKinstry, Renwick AG, Dirnhuber M, Waller DG, George CF, 1993. Quercetin, an in vitro inhibitor of CYP3A, does not contribute to the interaction between nifedipine and grapefruit juice. Br J Clin Pharmacol. 36:460–463.

Rashid TJ, Martin U, Clarke H, Waller DG, Renwick AG, George CF, 1995. Factors affecting the absolute bioavailability of nifedipine. Br J Clin Pharmacol. 40:51–58.

Rath K, Taxis K, Walz G, Gleiter C, Li S, Heide L, 2004. Pharmacokinetic Study of Artemisinin after oral intake of a traditional preparation of Artemisinin Annua L. (Annua Wormwood). Am J Trop Med Hyg. 70:128e132.

Ratheesh M, Shyni GL, Sindhu G, Helen A, 2011. Inhibitory effect of Ruta graveolens L. on oxidative damage, inflammation and aortic pathology in hypercholesteromic rats. Exp. Toxicol. Pathol. 63, 285–290.

Raza H, Ahmed I, John A, Sharma AK, 2000. Modulation of xenobiotic metabolism and oxidative stress in chronic streptozotocin-induced diabetic rats fed with Momordica charantia fruit extract. J. Biochem. Mol. Toxicol. 14, 131–139.

Raza H, Ahmed I, Lakhani MS, Sharma AK, Pallot D, Montague W, 1996. Effect of bitter melon (Momordica charantia) fruit juice on the hepatic cytochrome P450-dependent monooxygenases and glutathione S-transferases in streptozotocin-induced diabetic rats. Biochem. Pharmacol. 52, 1639–1642.

Razali N, Razab R, Junit SM, Aziz AA, 2008. Radical scavenging and reducing properties of extracts of cashew shoots (*Anacardium occidentale*). Food Chemistry 111(1), 38–44.

Rea AI, Schmidt JM, Setzer WN, Sibanda S, Taylor C, Gwebu ET, 2003. Cytotoxic activity of *Ozoroa insignis* from Zimbabwe. Fitoterapia 74(7), 732– 735.

Reddy PA, Rao JV, 2013. Evaluation of Anti-Inflammatory activity of *Evolvulus alsinoides* plant extracts. Journal of Pharmaceutical and Scientific Innovation 2(3), 24–26.

Rich DS, 2004. New JCAHO medication management standards for 2004. Am J Health Syst Pharm 61, 1349–1358.

Riddle JM1. History as a tool in identifying "new" old drugs. Adv Exp Med Biol. 505, 89-94.

Riet-correa F, Medeiros MA De, Pereira F, Dantas M, Maria R, Medeiros T, 2014. Administration of different concentrations of leaves of *Ipomoea asarifolia* in feed of mice. Ciencia Rural, Santa Maria 44(5), 872–877.

Rodeiro I, Donato MT, Martínez I, Hernández I, Garrido G, González-Lavaut JA, … Gómez-Lechón MJ, 2008. Potential hepatoprotective effects of new Cuban natural products in rat hepatocytes culture. Toxicology in Vitro 22(5), 1242–1249. http://doi.org/10.1016/j.tiv.2008.04.006

Rodríguez Cabrera D, 2003. El Mural de los Bebedores de Cholula, Puebla, Arqueología Mexicana, Vol. X, No. 59, enero-febrero, INAH, México, pp. 32-37.

Rogerio AP, Fontanari C, Borducchi E, Keller AC, Russo M, Soares EG, Albuquerque DA, Faccioli LH, 2008. Anti-inflammatory effects of Lafoensia pacari and ellagic acid in a murine model of asthma. European Journal of Pharmacology 580, 262-270.

Rong S, Zhao Y, Bao W, Xiao X, Wang D, Nussler AK, *et al.*, 2012. Curcumin prevents chronic alcohol-induced liver disease involving decreasing ROS generation and enhancing antioxidative capacity. Phytomedicine 19, 545–550.

Rosales D, Historia General del Reino de Chile, Flandes Indiano. Andrés Bello, Santiago, Chile. 1989.

Rose KD, Croissant PD, Parliament CF, Levin MB, 1990. Spontaneous spinal epidural hematoma with associated platelet dysfunction from excessive garlic ingestion: a case report. *Neurosurgery 26*(5), 880–2.

Rossier MF, Python CP, Capponi AM, Schlegel W, Kwan CY, Vallotton MB, 1993. Blocking T-type calcium channels with tetrandrine inhibits steroidogenesis in bovine adrenal glomerulosa cells. *Endocrinology 132*(3), 1035–43.

Roth BL *et al.*, 2004. Screening the receptorome to discover the molecular targets for plant-derived psychoactive compounds: a novel approach for CNS drug discovery. Pharmacol Ther. 102, 99-110.

Roth BL *et al.*, 2002. Salvinorin A: a potent naturally occurring nonnitrogenous kappa opioid selective agonist. Proc Natl Acad Sci USA 99, 11934 – 11939.

Rothman RB *et al.*, 2003. In vitro characterization of ephedrine-related stereoisomers at biogenic amine transporters and the receptorome reveals selective actions as norepinephrine transporter substrates. J Pharmacol Exp Ther. 307, 138-45.

Rowe RC *et al.*, 2009. Handbook of pharmaceutical excipients, 2nd ed: preface X. Washington (DC): American Pharmaceutical Association.

Roux S *et al.*, 2005. Primary observation (Irwin) test in rodents for assessing acute toxicity of a test agent and its effects on behavior and physiological function. Curr Protoc Pharmacol. Chapter 10: doi: 10.1002/0471141755.ph1010s27.

Ruck CAP, Bigwood J, Staples D, Ott J, and Wasson RG, 1979. 'Entheogens', Journal of Psychedelic Drugs, 11 (1–2), 145–6.

Russell WMS and Burch RL, The Principles of Humane Experimental Technique. London: Methuen and Co. [Reissued: 1992, Universities Federation for Animal Welfare, Herts, UK]; 1959.

Sá LZCM, Castro PFS, Lino FMA, Bernardes MJC, Viegas JCJ, Dinis TCP, Gil ES, 2014. Antioxidant potential and vasodilatory activity of fermented beverages of jabuticaba berry (Myrciaria jaboticaba). *Journal of Functional Foods 8*(1), 169–179.

Sadeghi Z, Kuhestani K, Abdollahi V, Mahmood A, 2014. Ethnopharmacological studies of indigenous medicinal plants of Saravan region, Baluchistan, Iran. *Journal of Ethnopharmacology 153*(1), 111–118.

Sadiq A, Hayat MQ, Ashraf M, *Artemisia annua* - Pharmacology and Biotechnology. In T. Aftab, J.F.S. Ferreira, and M.M.A. Khan (Eds.), Artemisia annua-Pharmacology and Biotechnology (pp. 9–25). Berlin Heidelberg. 2014.

Safadi R, Levy I, Amitai Y, Caraco Y, 1995. Beneficial effect of digoxin-specific Fab antibody fragments in oleander intoxication. *Archives of Internal Medicine 155*(19), 2121–5.

Saidu Y, Bilbis LS, Lawal M, Isezuo SA, Hassan SW, Abbas AY, 2007. Acute and sub-chronic toxicity studies of crude aqueous extract of *Albizia chevalieri* Harms (Leguminosae). Asian Journal of Biochemistry 2(4), 224–236.

Saha J, Mitra T, Gupta K, and Mukherjee S, 2012. Phytoconstituents and HPTLC analysis in Saraca asoca (roxb.) Wilde. Int. J. Pharm. Pharm. Sci. 4, 96–99.

Saha JC, Savini EC, and Kasinathan S, 1961. Ecbolic properties of Indian medicinal plants, part I. Indian J. Med. Res. 49, 130–151.

Saini ML, Saini R, Roy S, Kumar A, 2008. Comparative pharmacognostical and antimicrobial studies of acacia species (Mimosaceae). Journal of Medicinal Plants Research 2(12), 378–386.

Salama SM, Abdulla MA, AlRashdi AS, Ismail S, Alkiyumi SS, Golbabapour S, 2013. Hepatoprotective effect of ethanolic extract of *Curcuma longa* on thioacetamide induced liver cirrhosis in rats. BMC Complement. Altern. Med. 13:56.

Salgueiro JB *et al.,* 1997. Anxiolytic Natural and Synthetic Flavonoid Ligands of the Central Benzodiazepine Receptor Have No Effect on Memory Tasks in Rats. Pharmacology Biochemistry and Behavior 58, 887–891.

Sallau AB, Njoku GC, Olabisi AR, Wurochekke AU, Abdulkadir AA, Isah S, Ibrahim S, 2005. Effect of *Guiera senegalensis* leaf extract on some *Echis carinatus* venom enzymes. Journal of Medical Science 5(4), 280–283.

Samson R, Ramachandran R, Le Jemtel TH, 2014. Systolic heart failure: knowledge gaps, misconceptions, and future directions. *The Ochsner Journal 14*(4), 569–75.

Sani D, Sanni S, Sandabe UK, Ngulde SI, 2009. Effect of intake of aqueous stem extract of *Anisopus mannii* on haematological parameters in rats. International Journal of Applied Research in Natural Products 2(3), 22–28.

Sani D, Sanni S, Sandabe UK, Ngulde SI, 2010. Toxicological studies on *Anisopus mannii* crude aqueous stem extract/: biochemical and histopathological effects in albino rats. International Journal of Pharmaceutical Sciences 2(1), 51–59.

Sanni S, Thilza IB, Talle M, Mohammed SA, Sanni FS, Okpoli LA, Saleh M, 2010. The effect of *Acacia nilotica* pob ethyl acetate fraction on induced diarrhea in albino rats, 3(8).

Satpal Singh, T. H. Anantha Krishna, Subban Kamalraj, Gini C. Kuriakose, Jinu Mathew Valayil and Chelliah Jayabaskaran, 2015. Phytomedicinal importance of Saraca asoca (Ashoka): an exciting past, an emerging present and a promising future. Current Science 109, 10.

Sarker S. D. and Nahar L. Chemistry for Pharmacy Students General, Organic and Natural Product Chemistry, John Wiley & Sons Ltd, England

Sarkiyayi S, Umaru HA, Onche HO, 2015. Anti-inflammatory and analgesic effect of *Leptadenia hastata* on albino rats. American Journal of Biochemistry 5(2), 35–41.

Schaeffer J, 2012. Using Natural Therapies to Treat Hypertension – Strong Evidence Shows Certain Nutrients May Help Treat Hypertension When Combined With Dietary and Lifestyle Modifications. *Today's Dietitian* 14(2), 38.

Scherl EJ, Kumar S, Warren RU, 2010. Review of the safety and efficacy of ustekinumab. Therapeutic Advances in Gastroenterology 3(5), 321–328.

Schultes RE, and Hofmann A, Plantas de los Dioses, Origen del uso de los alucinógenos, Fondo de Cultura Económica, México. 2012.

Schultes RE, and Hofmann A, The Botany and Chemistry of Hallucinogens, Charles C. Thomas Publ., Springfield, Illinois. 1980.

Schulz KF, Altman DG, Moher D, the CONSORT Group (2010). CONSORT 2010 Statement: updated guidelines for reporting parallel group randomised trials. Br Med J 340: c332. doi: 10.1136/bmj.c332.

Schüssler M, Hölzl J, Fricke U, 1995. Myocardial effects of flavonoids from Crataegus species. *Arzneimittel-Forschung* 45(8), 842–5.

Senkoro AM, Barbosa FMA, Moiane SF, Albano G, Barros AIR, De. 2014. Bark stripping from forest tree species in Madjadjane, Southern Mozambique/: medicinal uses and implications for conservation. Natural Resources 5, 192–199.

Shah BH, Safdar B, Virani SS, Nawaz Z, Saeed SA, Gilani AH, 1997. The antiplatelet aggregatory activity of *Acacia nilotica* is due to blockade of calcium influx through membrane calcium channels. General Pharmacology 29(2), 251–255.

Shah SRU, Hassan G, Rehman A, Ahmad I, 2006. Ethnobotanical studies on the flora of district Muskhel and Barkhan in Balochistan, Pakistan. Pakistan Journal of Weed Science Research 12(3), 199–111.

Shah V, Sanmukhani J, 2010. Five cases of Jatropha curcas poisoning. The Journal of the Association of Physicians of India 58(April), 245–246.

Shanon B, 2002. Entheogens, Reflections on Psychoactive Sacramentals, Journal of Consciousness Studies 9(4), 85-94.

Shanthi S, Parasakthy K, Deepalakshmi PD, Devaraj SN, 1994. Hypolipidemic activity of tincture of Crataegus in rats. *Indian Journal of Biochemistry and Biophysics 31*(2), 143–6.

Sharma A, Sharma MK, Kumar M, 2007. Protective effect of *Mentha piperita* against arsenic-induced toxicity in liver of Swiss albino mice. Basic Clin. Pharmacol. Toxicol. 100, 249–257.

Sharma PV, Classical Uses of Medicinal Plants, Chaukhambha Visvabharati Publishers, Varanasi, India. 1996.

Sharma S, Dhamija HK, Parashar B, 2009. *Jatropha curcas*: a review. Asian Journal of Research in Pharmaceutical Sciences 2(3), 107–111.

Shaw D, Pearn J, 1979. Oleander poisoning. *The Medical Journal of Australia 2*(5), 267–9.

Shepard GH Jr, Psychoactive botanicals in ritual, religion and shamanism. In: *Ethnopharmacology*, E. Elisabetsky and N. Etkin (Eds.). Encyclopedia of Life Support Systems (EOLSS), Theme 6.79. Oxford, UK: UNESCO/Eolss Publishers. 2005.

Shi JS, Liu GX, Wu Q, Huang YP, Zhang XD, 1992. Effects of rhynchophylline and isorhynchophylline on blood pressure and blood flow of organs in anesthetized dogs. *Zhongguo Yao Li Xue Bao = Acta Pharmacologica Sinica 13*(1), 35–8.

Shirolkar A, Gahlaut A, Chhillar AK, and Dabur R, 2013. Quantitative analysis of catechins in Saracaasoca and correlation with antimicrobial activity. J. Pharm. Anal. 3, 421–428.

Shon YH, Nam KS, 2004. Inhibition of cytochrome P450 isozymes and ornithine decarboxylase activities by polysaccharides from soybeans fermented with Phellinus igniarius or Agrocybe cylindracea. Biotechnol. Lett. 26, 159–163.

Sibandze GF, van Zyl RL, van Vuuren SF, 2010. The anti-diarrhoeal properties of *Breonadia salicina*, *Syzygium cordatum* and *Ozoroa sphaerocarpa* when used in combination in Swazi traditional medicine. Journal of Ethnopharmacology 132(2), 506–511.

Siddhuraju P, Becker K, 2003. Antioxidant properties of various solvent extracts of total phenolic constituents from three different agroclimatic origins of drumstick tree (*Moringa oleifera* Lam.) leaves. Journal of Agricultural and Food Chemistry 51(8), 2144–2155.

Sides H, 2015. Marihuana Entre la Ciencia y el Prejuicio, National Geographic Magazine, pp. 3-29.

Sies, H, and Ketterer B, Glutathione Conjugation. Cornwall: Academic Press. 1988.

Sigush H, Hippius M, Henschel L, Kaufmann K, Hoffmann A, 1994. Influence of grapefruit juice on the pharmacokinetics of a slow release nifedipine formulation. Pharmazie. 49:522–524.

Singh A, 2008. Review of Ethnomedicinal uses and pharmacology of Evolvulus alsinoides Linn. Ethnobotanical Leaflets 12, 734–740.

Singh B, Kale RK, Rao AR, 2004. Modulation of antioxidant potential in liver of mice by kernel oil of cashew nut (Anacardium occidentale) and its lack of tumour promoting ability in DMBA induced skin papillomagenesis. Indian J. Exp. Biol. 42, 373–377.

Singh M, Khatoon S, Singh S, Kumar V, Rawat AKS, Mehrotra S. 2010. Antimicrobial screening of ethnobotanically important stem bark of medicinal plants. Pharmacognosy Research 2(4), 254–257.

Singh RK, Singh D, Mahendrakar AG, 2010. *Jatropha* poisoning in children. Medical Journal Armed Forces India 66(1), 80–81.

Singh S, Wahajuddin, Tewari D, Patel K, Jain G, 2011. Permeability determination and

pharmacokinetic study of nobiletin in rat plasma and brain by validated highperformance liquid chromatography method. Fitoterapia. 82:1206e1214.

Siripurapu KB, Gupta P, Bhatia G, Maurya R, Nath C, Palit G, 2005. Adaptogenic and anti-amnesic properties of *Evolvulus alsinoides* in rodents. Pharmacology Biochemistry and Behavior 81(3), 424–432.

Sloley BD *et al.,* 2000. Identification of kaempferol as a monoamine oxidase inhibitor and potential Neuroprotectant in extracts of Ginkgo biloba leaves. J Pharm Pharmacol. 52, 451-9.

Smolinske SC. Handbook of food, drug and cosmetic excipients, 1st ed. Boca Raton (FL): CRC Press; 1992.

Sobiecki JF, 2014. Psychoactive Plants: A Neglected Area of Ethnobotanical Research in Southern Africa (Review). Ethno Med 8(2), 165-172.

Soladoye MO, Adetayo MO, Chukwuma EC, Adetunji AN, 2014. Ethnobotanical survey of plants used in the treatment of female infertility in South-Western Nigeria. Ethnbotany Research and Application 12, 81–90.

Solon S, Lopes L, Sousa JR PT, Schmeda-Hirschmann G, 2000. Free radical scavenging activity of *Lafoensia pacari.* Journal of Ethnopharmacology 72, 173-178.

Somania R, Vadnala P, Deshmukhb S, Patwardhana S, 2011. Hepatoprotective activity of Clitoria ternatea L. leaves against carbon tetrachloride induced hepatic damage in rats. Journal of Pharmacy Research 4(10): 3540-3544.

Sombie PA, Hilou A., Mounier C., Coulibaly AY., Kiendrebeogo M., 2011. Antioxidant and anti-inflammaory activities from galls of *Gueira senegalensis* J.F. Gmel (Combretaceae). Research Journal of Medicinal Plants, 5(4), 448–461.

Sonibare MA, Gbile ZO, 2008. Ethnobotanical survey of anti-asthmatic plants in South Western Nigeria. African Journal of Traditional, Complementary, and Alternative Medicines 5(4), 340–345.

Soons PA, Vogels BAPM, Roosemalen MCM, *et al.,* 1991. Grapefruit juice and cimetidine inhibit stereoselective metabolism of nitrendipine in man. Clin Pharmacol Ther. 50:394–403.

Souza CD, Felfili JM, 2006. Uso de plantas medicinais na região de Alto Paraíso de Goiás, GO, Brasil. Acta Botânica Brasileira 20(1), 135-142.

Sowemimo A, van de Venter M, Baatjies L, Koekemoer T, 2009. Cytotoxic activity of selected Nigerian plants. African Journal of Traditional, Complementary and Alternative Medicines 6(4), 526–528.

Spadaro F, Costa R, Circosta C, Occhiuto F, 2012. Volatile composition and biological activity of key lime *Citrus aurantifolia* essential oil. Natural Product Communications 7(11), 1523–1526.

Steru L, *et al.*, 1985. The Tail Suspension Test; A New Method for Screening Antidepressants in Mice, Psychopharmacology 85, 367-370.

Strekalova T, *et al.*, 2004. Stress-Induced Anhedonia in Mice is Associated with Deficits in Forced Swimming and Exploration. Neuropsychopharmacology 29, 2007–2017.

Subehan Zaidi SF., Kadota S., Tezuka Y. (2007). Inhibition on human liver cytochrome P450 3A4 by constituents of fennel (Foeniculum vulgare): identification and characterization of a mechanism-based inactivator. J. Agric. Food Chem. 55, 10162–10167.

Sulaiman FA, Kazeem MO, Waheed AM, Temowo SO, Azeez IO, Zubair FI, ... Adeyemi OS, 2014. Antimicrobial and toxic potential of aqueous extracts of *Allium sativum*, *Hibiscus sabdariffa* and *Zingiber officinale* in Wistar rats. Journal of Taibah University for Science 8(4), 315–322.

Suleiman M, Shinkafi BY, Yusuf SH, 2014. Bioefficacy of leaf and peel extracts of *Euphorbia balsamifera* L. and *Citrus sinensis* L. against *Callosobruchus maculatus* Fab. [ Coleoptera/: Bruchidae ]. Annals of Biological Research 5(4), 6–10.

Sultana B, Anwar F, Przybylski R, 2007. Antioxidant activity of phenolic components present in barks of *Azadirachta indica*, *Terminalia arjuna*, *Acacia nilotica*, and *Eugenia jambolana* Lam. trees. Food Chemistry 104(3), 1106–1114.

Sultana B, Hussain Z, Asif M, Munir A, 2012. Investigation on the antioxidant activity of leaves, peels, stems bark, and kernel of mango (*Mangifera indica* L.). Journal of Food Science 77(8), 849–C852.

Summerfield A, Gunther MK, Thomas C, Mettenleiter B, Hanns-Joachim R, Armin S, 1997. Antiviral activity of an extract from leaves of the tropical plant *Acanthospermum hispidum*. Antiviral Research. 36, 55–62.

Sutter MC, Wang YX, 1993. Recent cardiovascular drugs from Chinese medicinal plants. *Cardiovascular Research* 27(11), 1891–901.

Swain AR, Dutton SP, Truswell AS, 1985. Salicylates in foods. *Journal of the American Dietetic Association* 85(8), 950–60.

Swamy TA, Jackie O, Mutuku NC, Africa E, 2013. *In vitro* control of selected pathogenic organisms by *Vernonia adoensis* roots. International Journal of Pharmacy and Life Sciences 4(8), 2855–2859.

Swamy TA, Obey J, Mutuku NC, 2013. Phytochemical analysis of *Vernonia adoensis* leaves and roots used as a traditional medicinal plants in Kenya. International Journal of Pharmacy and Biological Science 3(3), 46–52.

Sy GY, Sarr A, Dièye AM, Faye B, 2004. Myorelaxant and antispasmodic effects of the aqueous extract of *Mitragyna inermis* barks on Wistar rat ileum. Fitoterapia 75(5), 447–450.

Tabuti JRS, Kukunda CB, Waako PJ, 2010. Medicinal plants used by traditional medicine practitioners in the treatment of tuberculosis and related ailments in Uganda. Journal of Ethnopharmacology 127(1), 130–136.

Tahir, A. El, Satti, G.M.H., Khalid, S.A. 1999. Antiplasmodial activity of selected Sudanese medicinal plants with emphasis on. Journal of Ethnopharmacology 64, 227–233.

Takao K, and Miyakawa T, 2006. Light/dark Transition Test for Mice. J Vis Exp. 1, 104.

Takegoshi K, Tohyama T, Okuda K, Suzuki K, Ohta G, 1986. A case of Venoplant-induced hepatic injury. *Gastroenterologia Japonica 21*(1), 62–5.

Tamboura PH, Bayala B, Lompo M, Guissou PIP, Sawadogo L, 2005. Ecological distribution, morphological characteristics and acute toxicity of aqueous extracts of *Holarrhena floribunda* (G. don) Durand and Schinz, *Leptadenia hastata* (Pers.) Decne and *Cassia sieberiana* (DC) used by veterinary healers in Burkina Faso. African Journal of Traditional, Complementary and Alternative Medicines 2(1), 13–24.

Tandon N, Roy M, Roy S, Gupta N, 2012. Protective effect of psidium guajava in arsenic-induced oxidative stress and cytological damage in rats. Toxicol. Int. 19, 245–249.

Tang JR, Shi L, 1991. The inhibitory effects of sodium paeonol sulfate on calcium influx in the cultured neonatal rat heart cells during calcium paradox. *Yao Xue Xue Bao = Acta Pharmaceutica Sinica 26*(3), 161–5.

Tanko Y, Daniel PA, Mohammed KA, Jimoh A, Yerima M, Mohammed A, 2013. Effect of ethanolic extract of *Caralluma diazielli* on serum lipid profiles on fructose induced diabetes in wistar rats. Annals of Biological Research 4(2), 162–166.

Tapsoba H, Deschamps JP, 2006. Use of medicinal plants for the treatment of oral diseases in Burkina Faso. Journal of Ethnopharmacology 104(1-2), 68–78.

Tédong L, Dzeufiet PDD, Dimo T, Asongalem EA, Sokeng SD, Flejou JF,. Kamtchouing P, 2007. Acute and subchronic toxicity of *Anacardium occidentale* Linn (Anacardiaceae) leaves hexane extract in mice. African Journal of Traditional, Complementary and Alternative Medicines 4(2), 140–147.

Teng CM, Yu SM., Chen CC, Huang YL, Huang TF, 1990. EDRF-release and Ca+(+)-channel blockade by magnolol, an antiplatelet agent isolated from Chinese herb Magnolia officinalis, in rat thoracic aorta. *Life Sciences 47*(13), 1153–61.

Terashima H, Kalala S, Malasi N, 1992. Ethnobotany of the lega in the tropical rain forest of Eastern Zaire/: Part Two, Zone De Walikale. African Study Monographs 19(November), 1–60.

Thackaberry EA *et al.*, 2010. Comprehensive investigation of hydroxypropyl methylcellulose, propylene glycol, polysorbate 80, and hydroxypropyl-β-cyclodextrin for use in general toxicology studies. Toxicol Sci 117, 485–492.

Thapliyal R, Deshpande SS, Maru GB, 2002. Mechanism(s) of turmeric-mediated protective effects against benzo(a)pyrene-derived DNA adducts. Cancer Lett. 175, 79–88.

Thatipelli RC, Yellu NR, 2014. Hepatoprotective activity of *Evolvulus alsinoides* Linn. on paracetamol induced rats. Journal of Pharmaceutical and Scientific Innovation 3(4), 392–396.

Theoduloz C, Rodríguez JA, Pertino M, Schmeda-Hirschmann G, 2009. Antiproliferative activity of the diterpenes jatrophone and jatropholone and their derivatives. Planta Medica 75(14), 1520–1522.

Thomas G, Lucas P, Capler NR, Tupper KW, and Martin G, 2013. Ayahuasca-assisted therapy for addiction: Results from a preliminary observational study in Canada. Curr Drug Abuse Rev. 6(1), 30-42.

Thomas SD, 2012. *Leptadenia hastata*: a review of its traditional uses and its pharmacological activity. Medicinal Chemistry 02(07), 148–150.

Thomson M, Al-Amin ZM, Al-Qattan KK, Shaban LH, Ali M, 2007. Anti-diabetic and hypolipidaemic properties of garlic (*Allium sativum*) in streptozotocin-induced diabetic rats. International Journal of Diabetes and Metabolism 15(3), 108–115.

Threlfall R, Asian Journal of Organic Chemistry. Wiley-VCH. 2013. 10.1002/chemv.201200119.

Tijjani MB, Auwal MS, Mairiga IA, Saka S, Shuaibu A, Hambagda M, 2012. Acute toxicity, phytochemical, hypoglycaemic and antidiabetic activity of the aqueous stem extract of Anisopus mannii (Asclepiadaceae) on wistar strain albino rats. Africa Journal of Animal and Biomedical Sciences 7(1), 50–58.

Timothy SY., Bomai, HI., Musa, AH. 2015. Acute and subchronic toxicity study of the aqueous and ethanolic extracts of *Mitragyna inermis* bark in albino rats. International Journal of Pharmacology and Toxiology 5(1), 24–32.

Timothy SY, Wazis CH, Helga BI, Maina A, Bomai HI, 2014. Anticonvulsant screening of the aqueous and ethanol extracts of *Mitragyna inermis* bark. International Journal of Pharmacy and Therapeutics 5(5), 358–363.

Togola A, Austarheim I, Theïs A, Diallo D, Paulsen BS, 2008. Ethnopharmacological uses of *Erythrina senegalensis*: a comparison of three areas in Mali, and a link between traditional knowledge and modern biological science. Journal Ethnobiology and Ethnomedicine 4(6).

Traore MS, Baldé MA, Diallo MST, Baldé ES, Diané S, Camara A, … Baldé AM 2013. Ethnobotanical survey on medicinal plants used by Guinean traditional healers in the treatment of malaria. Journal of Ethnopharmacology 150(3), 1145–53.

Traore F, Gasquet M, Laget M, Guiraud H, Di Giorgio C, Azas N, ... Timon-David P, 2000. Toxicity and genotoxicity of antimalarial alkaloid rich extracts derived from *Mitragyna inermis* O. Kuntze and *Nauclea latifolia*. Phytotherapy Research 14(8), 608–611.

Tripoli E, Guardia M. La, Giammanco S, Majo DD, Giammanco M, 2007. *Citrus* flavonoids: Molecular structure, biological activity and nutritional properties: A review. Food Chemistry 104(2), 466–479.

Tucakov J, 1964. Pharmacognosy. Beograd: Institute for text book issuing in SR. Srbije; 11–30.

Turner PV, Brabb T, Pekow C, Vasbinder MA, 2011. Administration of substances to laboratory animals: routes of administration and factors to consider. *Journal of the American Association for Laboratory Animal Science/: JAALAS 50*(5), 600–13.

Turner PV et al., 2011, Administration of Substances to Laboratory Animals: Equipment Considerations, Vehicle Selection, and Solute Preparation. Journal of the American Association for Laboratory Animal Science/: JAALAS. 50, 614-627.

Tyler VE, The Honest Herbal: A Sensible Guide to the Use of Herbs and Related Remedies. *Pharmaceutical Product Press*. New York. 1993.

Tyler VE, *Herbs of Choice: The Therapeutic Use of Phytomedicinals. Pharmaceutical Product Press*. New York; 1994.

Tyler VE, *The honest herbal/: a sensible guide to the use of herbs and related remedies*. Pharmaceutical Products Press. New York; 1992.

Uche FI, Aprioku JS, 2008. The Phytochemical Constituents, Analgesic and Anti-inflammatory effects of methanol extract of *Jatropha curcas* leaves in Mice and Wistar albino rats. Journal of Applied Sciences and Environmental Management 12(4).

Ugwah-Oguejiofor CJ, Abubakar K, Ugwah MO, Njan AA, 2013. Evaluation of the antinociceptive and anti-inflammatory effect of *Caralluma dalzielii*. Journal of Ethnopharmacology 150(3), 967–972. http://doi.org/10.1016/j.jep.2013.09.049

Ujah OF, Ujah IR, Johnson JT, Oka VO, 2013. Effect of ethanolic leaf extract of *Moringa olifera* leaf on haematological and biochemical parameters of wistar rats. Journal of Natural Product and Plant Resources 3(2), 10–14.

Umaru HA, Shugaba A, Addy EO, 2014. Effect of *Leptadenia hastata* (Pers.) Decne on metabolic profile of pregnant albino rats. Asian Journal of Medical Sciences 6(3), 30–33.

Unger M, Frank A, 2004. Simultaneous determination of the inhibitory potency of herbal extracts on the activity of six major cytochrome P450 enzymes using liquid chromatography/mass spectrometry and automated online extraction. Rapid Commun. Mass Spectrom. 18, 2273–2281.

UNICEF/UNDP/World Bank/WHO, Special Programme for Research and Training in Tropical Diseases (TDR) Handbook Non-Clinical Safety Testing. Avenue Appia 20, 1211 Geneva 27 – Switzerland; 2012.

Uno T, Ohkubo T, Sugawara K, Higashiyama A, Motomura Motomura. 1997. Effect of grapefruit juice on the disposition of nicardipine after administration of intravenous and oral doses. Clin Pharmacol Ther. 61:209.

Usia T, Iwata H, Hiratsuka A, Watabe T, Kadota S, Tezuka Y, 2006. CYP3A4 and CYP2D6 inhibitory activities of Indonesian medicinal plants. Phytomedicine 13, 67–73.

Vajha M, Chillara SRK, 2014. Evaluation of cellular antioxidant activity of selected species of Caralluma and Boucerosia on cell lines. International Journal of Applied Sciences and Biotechnology 2(1), 83–87.

Van der Kooy F, Sullivan SE, 2013. The complexity of medicinal plants: the traditional Artemisia annua formulation, current status and future perspectives. Journal of Ethnopharmacology 150(1), 1–13.

Vanachayangkul P, Butterweck V, Frye R, 2009. Determination of Visnagin in rat plasma by liquid chromatography with tandem mass spectrometry and its application to in vivo pharmacokinetic studies. J Chromatogr B. 877:653e656.

VanPool CS, 2003. Viajes Chamánicos, Iconografía de Casas Grandes, Arqueología Mexicana, Vol. X, No. 59, enero-febrero, INAH, México, pp. 42-43.

Ventrella E, Marciniak P, Adamski Z, Rosiñski G, Chowañski S, Falabella P, Bufo AS, 2015. Cardioactive properties of Solanaceae plant extracts and pure glycoalkaloids on Zophobas atratus. *Insect Science* 22(2), 251–62.

Verma AR, Vijayakumar M, Mathela CS, Rao CV, 2009. *In vitro* and *in vivo* antioxidant properties of different fractions of *Moringa oleifera* leaves. Food and Chemical Toxicology 47(9), 2196–2201.

Vijaya Kumar S, Sankar P, Varatharajan R, 2009. Pharmaceutical Biology 47(10), 973 -975.

Vincent J, Foulds G, Dogolo LC, Willavize SA, Friedman HL, 1997. Grapefruit juice does not alter the pharmacokinetics of amlodipine in man (Abstract) Clin Pharmacol Ther. 61:233.

Voigt E, Junger H, 1978. Acute posttraumatic renal failure following therapy with antibiotics and beta-aescin [in German]. Anaesthesist 2781– 83.

Volpato G, Godínez D, Beyra A, Barreto A, 2009. Uses of medicinal plants by Haitian immigrants and their descendants in the Province of Camagüey, Cuba. Journal of Ethnobiology and Ethnomedicine 5, 16.

Walf AA and Frye CA, 2007. The use of the elevated plus maze as an assay of anxiety-related behavior in rodents. Nature Protocols 2, 322–328.

Wälli F, Grob PJ, Müller-Schoop J, 1981. Pseudo-(venocuran-)lupus—a minor episode in the history of medicine. *Schweizerische Medizinische Wochenschrift* 111(38), 1398–405.

Wang T, Liu Z, Li J, *et al.*, 2007. Determination of protodioscin in rat plasma by liquid chromatography-tandem mass spectrometry. J Chromatogr B. 848: 363e368.

Wang Z, Roberts JM, Grant PG, Colman RW, Schreiber AD, 1982. The effect of a medicinal Chinese herb on platelet function. Thrombosis and Haemostasis *48*(3), 301–6.

Wasowski C *et al.*, 2002. Isolation and identification of 6-methylapigenin, a competitive ligand for the brain GABAA receptors, from Valeriana wallichii. Planta Med 68, 934–936.

Weathers PJ, Towler M, Hassanali A, Lutgen P, Engeu PO, 2014. Dried-leaf of *Artemisia annua*: A practical malaria therapeutic for developing countries? World Journal of Pharmacology 3(4), 39.

Weiner ML and Kotkoskie LA. Excipient toxicology and safety, 1st ed. New York (NY): Marcel Dekker; 2000.

Wender PA, Kee JM, Warrington JM, 2008. Practical synthesis of prostratin, DPP, and their analogs, adjuvant leads against latent HIV. Science 320(5876), 649–652.

Wieraszko A *et al.*, 1993. Long-term potentiation in the hippocampus induced by platelet-activating factor. Neuron 10, 553- 557.

Willcox M, 2009. *Artemisia* species: from traditional medicines to modern antimalarials-and back again. The Journal of Alternative and Complementary Medicine 15(2), 101–109.

Willcox M, Bodeker G, Bourdy G, Dhingra V, Falquet J, Ferreira JFS, … Provendier D, 2004. *Artemisia annua* as a traditional herbal antimalarial. Traditional Medicinal Plants and Malaria 43–60.

Williams ET, Barminas JT, Akinniyi J, William A, 2009. Antidiarrhoeal effects of the root extracts of *Guiera senegalensis* in male mice. Pure and Applied Chemistry 3(8), 152–157.

Williamson, EM, Driver, S, and Baxter K, Stockley's Herbal Medicines Interactions: A Guide to the Interactions of Herbal Medicines, Dietary Supplements and Nutraceuticals with Conventional Medicines. London; Chicago: PharmaceuticalPress. 2009.

Williamson EM *et al.*, Pharmacological Methods in Phytotherapy Research V1. John Wiley and Sons Ltd; West Sussex; 1996, 15–23.

Willner P, 1987. Reduction of sucrose preference by chronic unpredictable mild stress, and its restoration by a tricyclic antidepressant. Psychopharmacology (Berl) 93, 358–364.

Wondimu T, Asfaw Z, Kelbessa E, 2007. Ethnobotanical study of medicinal plants around Dheeraa town, Arsi Zone, Ethiopia. Journal of Ethnopharmacology 112, 152–161. http://doi.org/10.1016/j.jep.2007.02.014

Wu X, Rao MR, 1990. Effects of praeruptorin C on isolated guinea pig atrium and myocardial compliance in patients. *Zhongguo Yao Li Xue Bao = Acta Pharmacologica Sinica 11*(3), 235–8.

Xavier TF, Kannan M, Auxilia A, 2015. Original Research Article Traditional Medicinal Plants Used in the treatment of different skin diseases. International Journal of Current Microbiology and Applied Sciences 4(5), 1043–1053.

Xia G, Yao W, Liu X, Jiang M, 1990. Effects of benzyltetrahydropalmatine on ischaemia-reperfusion with monophasic action potential. *Yao Hsueh Hsueh Pao 10*, 1–4.

Yadav NK *et al.*, 2015. Saraca indica bark extract shows in vitro antioxidant, antibreast cancer activity and does not exhibit toxicological effects. Oxid. Med. Cell. Longev. doi: 10.1155/2015/205360.

Yaghmaie P, Parivar K, Haftsavar M, 2011. Effects of *Citrus aurantifolia* peel essential oil on serum cholesterol levels in Wistar rats. Journal of Paramedical Sciences 2(1), 29–32.

Yagi A, Fujimoto K, Tanonaka K, Hirai K, Takeo S, 1989. Possible active components of tan-shen (Salvia miltiorrhiza) for protection of the myocardium against ischemia-induced derangements. *Planta Medica 55*(1), 51–4.

Yao JA, Zhang BH, 1989. Anti-arrhythmic mechanisms of sophoramine. *Zhongguo Yao Li Xue Bao = Acta Pharmacologica Sinica 10*(4), 315–9.

Yao WX, Xia GJ, Zhang JS, Zeng WZ, Zhang SD, Jiang MX, 1990. A new kalium channel blocker of Chinese medicinal origin—benzyltetrahydropalmatine hydrochloride. *Journal of Tongji Medical University = Tong Ji Yi Ke Da Xue Xue Bao 10*(1), 1–4.

Yee GC, Stanley DL, Pessa JL, Costa TD, Beltz SE, Ruiz J, Lowenthal DT, 1995. Effect of grapefruit juice on blood cyclosporin concentration. Lancet 345:955–956.

Yeung KS, Gubili J, Cassileth BR, 2011. Evidence-based Anticancer. Materia Medica. Media. http://doi.org/10.1007/978-94-007-0526-5

Yu SM, Chen CC, Huang YL, Tsai CW, Lin CH, Huang TF, Teng CM, 1990. Vasorelaxing effect in rat thoracic aorta caused by denudatin B, isolated from the Chinese herb, magnolia fargesii. European Journal of Pharmacology 187(1), 39–47.

Yusuf OS, Maxwell E, 2010. Analgesic activity of the methanolic leaf extract of Jatropha curcas (Linn) in. African Journal of Biomedical Research 13(May), 149–152.

Yusuf S, Bilbs LS, Lawal M, Isezu SA, Umar RA, 2007. Hematotoxicity study of the leaf extract of *Albizia chevaleiri* Harms (Leguminosae). Biochemia Medica 17(2), 64–74.

Z'Brun, 1995. A Ginkgo: myth and reality [in German]. *Schweiz Rundsch Med Prax.* 841–846.

Zadelaar S, Kleemann R, Verschuren L, Weij JVV, Hoorn JV, Princen HM, Kooistra T, 2007. Mouse models for atherosclerosis and pharmaceutical modifiers. *Arteriosclerosis, Thrombosis, and Vascular Biology 27*(8), 1706–21.

Zaragoza C, Gomez-Guerrero C, Martin-Ventura JL, Blanco-Colio L, Lavin B, Mallavia B, Egido J, 2011. Animal models of cardiovascular diseases. Journal of Biomedicine and Biotechnology 2011, 497841.

Zaruwa MZ, Manosroi A, Akihisa T, Manosroi W, Rangdaeng S, Manosroi J, 2012. Hypoglycemic activity of the Anisopus mannii NE Br. methanolic leaf extract in normal and alloxan-induced diabetic mice. Journal of Complementary and Integrative Medicine 10(1), 37–46.

Zhang W, Wojta J, Binder BR, 1994. Effect of notoginsenoside R1 on the synthesis of tissue-type plasminogen activator and plasminogen activator inhibitor-1 in cultured human umbilical vein endothelial cells. Arteriosclerosis and Thrombosis/: A Journal of Vascular Biology/American Heart Association 14(7), 1040–6.

Zhang ZG, Lu XB, Xiao L, Tang L, Zhang LJ, Zhang T, et al., 2012. Antioxidant effects of the Uygur herb, Foeniculum Vulgare Mill, in a rat model of hepatic fibrosis. Zhonghua Gan Zang Bing Za Zhi 20, 221–226.

Zhou, SF, Liu JP, and Chowbay B, 2009. Polymorphism of human Cytochrome P450 enzymes and its clinical impact. Drug Metab. Rev. 41, 89–295.

Zhou S, et al., 2003. Interactions of herbs with cytochrome P450. Drug Metab Rev. 35, 35-98.

Zhu JQ, Zeng FD, Hu CJ, 1990. Effects of dauricine on His-bundle electrogram and interaction with other drugs in anesthetized rabbits. Yao Xue Xue Bao = Acta Pharmaceutica Sinica 25(1), 6–10.

Ziblim IA, Timothy KA, Deo-anyi EJ, 2013. Exploitation and use of medicinal plants, Northern region, Ghana. Journal of Medicinal Plants Research 7(27), 1984–1993.

Zongo C, Akomo EFO, Savadogo A, Obame LC, Koudou J, Traore AS, 2009. In vitro antibacterial properties of total alkaloids extract from mitragyna inermis (Willd.) O. Kuntze, a West African traditional medicinal plant. Asian Journal of Plant Sciences 8(2), 172–177.

# Index

www.ingramcontent.com/pod-product-compliance
Lightning Source LLC
Chambersburg PA
CBHW050515190326
41458CB00005B/1547

* 9 7 8 1 7 8 7 1 5 0 0 0 3 *